PE Civil Structural Depth Six-Minute Problems

Eighth Edition

Christine A. Subasic, PE

PPI®
PPI2PASS.COM
A KAPLAN COMPANY

Register Your Book at ppi2pass.com

- Receive the latest exam news.
- Obtain exclusive exam tips and strategies.
- Receive special discounts.

Report Errors for This Book

PPI is grateful to every reader who notifies us of a possible error. Your feedback allows us to improve the quality and accuracy of our products. Report errata at **ppi2pass.com**.

PE CIVIL STRUCTURAL DEPTH SIX-MINUTE PROBLEMS
Eighth Edition

Current release of this edition: 1

Release History

date	edition number	revision number	update
Dec 2017	7	1	New edition. Code update. New content. Copyright update.
Sep 2018	7	2	Minor corrections. Minor cover updates.
Jun 2019	8	1	New edition. Code update. Copyright update.

Printed in the United States of America.

PPI
1250 Fifth Avenue, Belmont, CA 94002
(650) 593-9119
ppi2pass.com

ISBN: 978-1-59126-667-9

Table of Contents

Preface, Dedication, and Acknowledgments

The topics covered in this eighth edition of *PE Civil Structural Depth Six-Minute Problems* correspond with the subject areas identified by the National Council of Examiners for Engineering and Surveying (NCEES) for the structural depth section of the PE Civil exam. All problems have been reviewed and updated to the most recent codes and exam specifications, and to ensure problems are reflective of the type found on the structural sections of the PE Civil exam.

I wrote the problems in this book to be both conceptual and practical, and they are written to provide varying levels of difficulty. Though you won't encounter problems on the exam exactly like those presented here, working these problems and reviewing the solutions will increase your familiarity with the exam problems' form, content, and solution methods. This preparation will help you considerably during the exam.

All problems and solutions have been carefully prepared and reviewed to ensure that they are appropriate and understandable and that they are solved correctly. If you find errors or discover an alternative, more efficient way to solve a problem, please bring it to PPI's attention so your suggestions can be incorporated into future editions. You can report errors and keep up with the changes made to this book by logging onto PPI's website at **ppi2pass.com**.

I would like to dedicate this book to Shawn, my husband and biggest supporter, without whom I could never have worked all the hours necessary to complete this project. I would also like to acknowledge the support of my friends and mentors in the industry, particularly Phillip Samblanet, who always encouraged me in my quest for balance between family and my engineering career, and Maribeth Bradfield, who got me started writing problems in the first place.

I am also indebted to Valoree Eikinas, Robert Macia, and all the engineers at Stewart Engineering who "tried out" the problems in this book and to Thomas H. Miller, PhD, PE, for his work in completing the technical review of the first edition. I would like to thank Alireza Sayah, PhD, PE, for his work in updating the concrete problems. I also recognize several individuals for their assistance in updating past editions. Specifically, I would like to thank Jennifer Tanner Eisenhauer, PhD, PE, for her assistance in updating the steel and concrete problems and Matt Yerkey, PE, for his contributions to the bridge problems. I would also like to thank Elaine Huang, PE, for contributing a construction problem to help round out coverage. Thanks also to Anil Acharya, PE, for checking calculations.

On PPI's staff, my thanks go to the PPI product team, including Megan Synnestvedt, senior product manager; Steve Shea, content specialist; Tyler Hayes, senior copy editor; Bradley Burch, production editor; Beth Christmas, editorial project manager; Tom Bergstrom, technical illustrator and cover designer; Richard Iriye, typesetter; Cathy Schrott, production services manager; and Grace Wong, director of editorial operations, for their support in producing this book. Lastly, I thank God for giving me the talents to pursue this endeavor.

Christine A. Subasic, PE

About the Author

Christine A. Subasic, PE, is a consulting architectural engineer licensed in North Carolina and Virginia. Ms. Subasic graduated with a bachelor of architectural engineering degree, structures option, with honors and high distinction, from The Pennsylvania State University. For more than 25 years, she has specialized in masonry and structural design, standards development, and technical writing. Ms. Subasic has design experience in both commercial and residential construction. Her clients include commercial businesses, as well as several trade associations.

Ms. Subasic is an active member of ASTM International and the ASHRAE 189.1 Standard Project Committee, organizations committed to the development of construction industry standards. Ms. Subasic has served on the board of directors for The Masonry Society, and she is currently President-Elect and active on their Executive Committee, Design Practice Committee, and Sustainability Committee.

Ms. Subasic is the author of numerous articles, papers, and technical publications. She has authored chapters in the *Masonry Designers' Guide*, published by The Masonry Society, and co-authored *An Investigation of the Effects of Hurricane Opal on Masonry.* She has served as a subject matter expert for revisions of timber and masonry material in PPI publications. Her articles on many aspects of masonry have appeared in the magazines *Masonry Construction, Masonry,* and *STRUCTURE.*

Introduction

ABOUT THIS BOOK

PE Civil Structural Depth Six-Minute Problems contains problems that correspond to the format and scope that would be expected to appear in the PE Civil exam's structural depth section.

Most of the problems are quantitative, requiring calculations to arrive at a correct solution, while a portion are non-quantitative, requiring an understanding of theory and principle. Some problems will require a little more than six minutes to answer, and others require a little less.

All problems include hints to provide direction in their solving approaches. The solutions are presented in a step-by-step sequence to help you follow the logical development of the solving approach and to provide examples of how you may want to solve similar problems on the exam.

Solutions also include explanations of the faulty solutions leading to the three incorrect answer options. The incorrect options represent answers found by making common mistakes. These may be simple mathematical errors, such as failing to square a term in an equation, or more serious errors, such as using the wrong equation.

The solution presented for each problem may represent only one of several methods for obtaining the correct answer. Although most of these problems have unique solutions, alternative problem-solving methods may produce different, but nonetheless appropriate, answers.

ABOUT THE EXAM

The Principles and Practice of Engineering (PE) exam for civil engineering, administered by the National Council of Examiners for Engineering and Surveying (NCEES), is created from problems developed by educators and professional engineers representing consulting, government, and industry. The PE exam is designed to test examinees' understanding of both conceptual and practical engineering concepts.

The PE Civil exam is divided into two four-hour sections, with each session containing 40 multiple-choice problems. Only one of the four options given is correct, and the problems are completely independent of each other. Problems from past exams are not available from NCEES or any other source. However, NCEES does identify the general subject areas covered on the exam.

The morning section of the PE Civil exam is a "breadth" exam covering eight areas of general civil engineering knowledge: project planning, means and methods, soil mechanics, structural mechanics, hydraulics and hydrology, geometrics, materials, and site development. All examinees take the same breadth exam.

For the afternoon section, you must choose one of the five depth modules: construction, geotechnical, structural, transportation, or water resources and environmental. The structural depth section is intended to assess your knowledge of structural engineering principles and practices. The topics and approximate distribution of problems for the structural depth section are as shown in Table 1. On average, during the exam you should expect to complete 80 problems in 480 minutes (8 hours), or spend 6 minutes per problem.

THIS BOOK'S ORGANIZATION

PE Civil Structural Depth Six-Minute Problems is organized into three chapters: Analysis of Structures (14 problems), Design and Details of Structures (20 problems), Codes and Construction (6 problems). Each of the problems is representative of those found on the structural depth section of the exam. The three chapters of the book are further subdivided into topic areas that correspond to the topics given in the NCEES specifications for the structural depth section.

HOW TO USE THIS BOOK

To optimize your study time and obtain the maximum benefit from the practice problems, consider the following suggestions.

1. Complete an overall review of the problems, and identify the subjects that you are least familiar with. Work a few of these problems to assess your general understanding of the subjects and to identify your strengths and weaknesses.

2. Locate and organize relevant resource materials. (See the Codes and References section of this book as a starting point.) As you work problems, some of these resources will emerge as more useful to you than others. These are what you will want to have on hand when taking the PE exam.

3. Work the problems in one subject area at a time, starting with the subject areas that you have the most difficulty with.

4. When possible, work problems without utilizing the hints. Always attempt your own solutions before looking at the solutions provided in the book. Use the solutions to check your work or to provide guidance in solving the more difficult problems. Use the incorrect solutions to help identify pitfalls and to develop strategies to avoid them.

5. Use each solution as a guide to understanding general problem-solving approaches. Although problems identical to those presented in *PE Civil Structural Depth Six-Minute Problems* will not be encountered on the PE exam, the approach to solving problems will be the same.

For further information and tips on how to prepare for the PE Civil exam's structural depth section, consult the *PE Civil Reference Manual* or PPI's website at **ppi2pass.com**.

Table 1 *Exam Specifications for the PE Civil Structural Depth Exam*

Analysis of Structures (14 questions)

Loads and load applications: dead loads; live loads; construction loads; wind loads; seismic loads; moving loads (e.g., vehicular, cranes); snow, rain, ice; impact loads; earth pressure and surcharge loads; load paths (e.g., lateral and vertical); load combinations; tributary areas

Forces and load effects: diagrams (e.g., shear and moment); axial (e.g., tension and compression); shear; flexure; deflection; special topics (e.g., torsion, buckling, fatigue, progressive collapse, thermal deformation, bearing)

Design and Details of Structures (20 questions)

Materials and material properties: concrete (e.g., plain, reinforced, cast-in-place, precast, pre-tensioned, post-tensioned); steel (e.g., structural, reinforcing, cold-formed); timber; masonry (e.g., brick veneer, CMU)

Component design and detailing: horizontal members (e.g., beams, slabs, diaphragms); vertical members (e.g., columns, bearing walls, shear walls); systems (e.g., trusses, braces, frames, composite construction); connections (e.g., bearing, bolted, welded, embedded, anchored); foundations (e.g., retaining walls, footings, combined footings, slabs, mats, piers, piles, caissons, drilled shafts)

Codes and Construction (6 questions)

Codes, standards, and guidance documents: International Building Code (IBC); American Concrete Institute (ACI 318, 530); Precast/Prestressed Concrete Institute (PCI Design Handbook); Steel Construction Manual (AISC); National Design Specification for Wood Construction (NDS); LRFD Bridge Design Specifications (AASHTO); Minimum Design Loads for Buildings and Other Structures (ASCE 7); American Welding Society (AWS D1.1, D1.2, and D1.4); OSHA 1910 General Industry and OSHA 1926 Construction Safety Standards

Temporary structures and other topics: special inspections; submittals; formwork; falsework and scaffolding; shoring and reshoring; concrete maturity and early strength evaluation; bracing; anchorage; OSHA regulations; safety management

Codes and References Used to Prepare This Book

The minimum recommended library for the PE Civil structural depth exam includes the NCEES-adopted design codes, the *PE Civil Reference Manual* and the *PE Civil Structural Depth Six-Minute Problems.* Most problems on the PE Civil structural depth exam can be solved using the NCEES-adopted design codes and your knowledge of general engineering principles. As a general rule, you shouldn't bring books to the exam that you didn't use during your exam review.

The information that was used to write this book was based on exam specifications at the time of publication. However, as with engineering practice itself, the PE Civil structural depth exam is not always based on the most current codes or cutting-edge technology. Similarly, codes, standards, and regulations adopted by state and local agencies often lag issuance by several years. It is likely that the codes you use in practice and the codes that are the basis of your exam will all be different.

PPI lists on its website the dates and editions of the codes, standards, and regulations on which NCEES has announced that the PE Civil structural depth exam is based (**ppi2pass.com**). It is your responsibility to find out which codes are relevant to the PE Civil structural depth exam. In the meantime, here are the codes and standards that have been incorporated into this edition.

NCEES CODES

AASHTO: *AASHTO LRFD Bridge Design Specifications*, 7th ed. (without interims), 2015, American Association of State Highway and Transportation Officials, Washington, DC.

ACI 318: *Building Code Requirements for Structural Concrete*, 2014, American Concrete Institute, Farmington Hills, MI.

AISC: *Steel Construction Manual,* 14th ed., 2011, American Institute of Steel Construction, Inc., Chicago, IL.

ASCE/SEI7: *Minimum Design Loads for Buildings and Other Structures*, 3rd printing, 2010, American Society of Civil Engineers, Reston, VA.

IBC: *International Building Code*, 2015 ed. (without supplements), International Code Council, Inc., Falls Church, VA.

NDS: *National Design Specification for Wood Construction*, 2015 ed., and *National Design Specification Supplement, Design Values for Wood Construction*, 2015 ed., American Wood Council, Leesburg, VA.

NDS: *Special Design Provisions for Wind and Seismic*, 2015 ed., American Wood Council, Leesburg, VA.

OSHA 29 CFR Parts 1910 and 1926: *Safety Standards for General Industry and Construction*, U.S. Federal Version, January 2016. U.S. Department of Labor, Washington, DC.

PCI: *PCI Design Handbook: Precast and Prestressed Concrete*, 7th ed., 2010, Precast/Prestressed Concrete Institute, Chicago, IL.

TMS 402/602: *Building Code Requirements and Specification for Masonry Structures* (and related commentaries), 2013, The Masonry Society, Longmont, CO.

REFERENCES AND OTHER CODES IN THIS BOOK

The following references were used to prepare this book. They may also be useful resources for exam preparation.

American Institute of Timber Construction. *Standard Specification for Structural Glued Laminated Timber of Softwood Species* (AITC 117).

ASTM International. *Standard Specification for Hollow Brick* (*Hollow Masonry Units Made from Clay or Shale*) (ASTM C652). ASTM International.

Bowles, Joseph E. *Foundation Analysis and Design.* New York: McGraw-Hill.

The Masonry Society. *Masonry Designers' Guide* (MDG). Longmont, CO: The Masonry Society.

McCormac, Jack C. *Structural Analysis.* New York: Harper & Row.

Nilson, Arthur H., David Darwin, and Charles W. Dolan. *Design of Concrete Structures.* New York: McGraw-Hill.

Nomenclature

a	dimension	ft	m
A	area	in^2, ft^2	mm^2, m^2
A	cross-sectional area	in^2, ft^2	mm^2, m^2
A_t	cross-sectional area of torsion reinforcement	in^2	mm^2
b	dimension	in	mm
b	dimension from web to centerline of bolt hole in hanger connection	in	mm
b	width	in	mm
b	width of footing	ft	m
b_a	tension per bolt	lbf, kips	N
b_e	effective width of slab	in	mm
b_v	shear per bolt	lbf, kips	N
b_x, b_y	bending coefficients	$(\text{ft-kips})^{-1}$	$(\text{N·m})^{-1}$
B	allowable tension per bolt	kips	N
B	width of column base plate in direction of column flange	in	mm
B	width of footing	ft	m
B_a	allowable tension per bolt	kips	N
B_v	allowable shear per bolt	lbf	N
c	neutral axis depth	in	mm
c	undrained shear strength (cohesion)	lbf/ft^2	Pa
c_A	adhesion	lbf/ft^2	Pa
C	correction factor	various	various
C_f	force coefficient	–	–
C_1	moment coefficient	–	–
C_D	drag coefficient	–	–
d	beam depth	in	mm
d	beam depth (to tension steel centroid)	in	mm
d	bolt diameter	in	mm
d	diameter	in, ft	mm, m
d	effective depth	in	mm
d'	diameter of bolt hole	in	mm
d_b	nominal diameter of reinforcing bar	in	mm
D	dead load	lbf, kips, lbf/ft^2	N, Pa
D	depth	ft, in	m
D	diameter	ft	m
DF	distribution factor	–	–
e	eccentricity	in, ft	mm, m

E	earthquake load	kips	N
E	modulus of elasticity	lbf/in^2, lbf/ft^2, $kips/in^2$	Pa
E'	allowable modulus of elasticity	lbf/in^2, lbf/ft^2, $kips/in^2$	Pa
f	stress	lbf/in^2	Pa
f_a	compressive stress in masonry due to axial load alone	lbf/in^2	Pa
f_b	bending stress	lbf/in^2	Pa
f_b	stress in masonry due to flexure alone	lbf/in^2	Pa
f'_c	compressive strength of concrete	lbf/in^2	Pa
f'_m	compressive strength of masonry	lbf/in^2	Pa
f_r	modulus of rupture	lbf/in^2	Pa
f_s	shear stress	lbf/in^2	Pa
f_s	stress in steel reinforcement	lbf/in^2	Pa
f_t	tensile stress	lbf/in^2	Pa
f_v	shear stress in masonry	lbf/in^2	Pa
f_y	yield stress	lbf/in^2	Pa
F	allowable stress	$kips/in^2$	Pa
F	factor of safety	–	–
F	force	lbf, kips	N
F	strength	$kips/in^2$	Pa
F'	reduced allowable stress	$kips/in^2$	Pa
F_a	allowable compressive stress due to axial load alone	lbf/in^2	Pa
F_b	allowable bending stress	lbf/in^2	Pa
F_b	allowable compressive stress due to flexure alone	lbf/in^2	Pa
F_p	allowable bearing pressure	lbf/in^2	Pa
F_s	allowable tensile stress in steel reinforcement	lbf/in^2	Pa
F_t	allowable tensile stress	lbf/in^2	Pa
F_{TH}	threshold stress range	$kips/in^2$	Pa
F_v	allowable shear stress in masonry	lbf/in^2	Pa
F_y	yield strength of steel	lbf/in^2	Pa
FEM	fixed-end moment	ft-kips	N·m
g	lateral gage spacing of adjacent holes	in	mm

g	ratio of distance between tension steel and compression steel to overall column depth	–	–
G	gust factor	–	–
h	distance	ft	m
h	effective height	in	mm
h	height	in, ft	mm, m
h	overall thickness	in, ft	mm, m
h'	modified overall thickness	in, ft	mm, m
H	distance from soil surface to footing base	ft	m
H_s	length of stud connector	in	mm
I	moment of inertia	in^4	mm^4
I_p	polar moment of inertia	in^4	mm^4
j	ratio of distance between centroid of flexural compressive forces and centroid of tensile forces to depth	–	–
k	coefficient	–	–
k	coefficient of lateral earth pressure	–	–
k	effective length factor	–	–
k	stiffness	lbf/ft	N/m
K	coefficient	–	–
K	effective length factor	–	–
K	relative stiffness	ft^{-1}	m^{-1}
l	distance between points of lateral support of compression member in a given plane	in, ft	mm, m
l	length	in, ft	mm, m
l	span length	in, ft	mm, m
l_b	effective embedment length of headed or bent anchor bolts	in	mm
l_c	vertical distance between supports	in	mm
l_d	development length of reinforcement	in	mm
l_h	distance from center of bolt hole to beam end	in	mm
l_n	clear span length	in, ft	mm, m
l_u	unsupported length of a compression member	in	m
l_v	distance from center of bolt hole to edge of web	in	mm
L	length	in, ft	m
L	length of footing	ft	m
L	live load	lbf, kips, lbf/ft^2	N, Pa
L	span length	in, ft	mm, m
M	moment	in-lbf, ft-lbf, ft-kips	N·mm, N·m
M_a	maximum moment	in-lbf	N·m
M_m	resisting moment assuming masonry governs	in-lbf	N·m
M_o	total factored static moment	in-lbf	N·m
M_R	beam resisting moment	ft-kips	N·mm
M_s	resisting moment assuming steel governs	in-lbf	N·m
M_t	torsional moment	ft-kips	N·mm
n	modular ratio	–	–
n	quantity (number of)	–	–
N	bearing capacity factor	–	–
N	length of column base plate in direction of column depth	in	mm
N_r	number of studs in one rib	–	–
p	perimeter	in, ft	mm, m
p	pressure	lbf/ft^2	Pa
P	axial load	kips	N
P	load	kips	N
P	prestress force in a tendon	lbf, kips	N
P_a	allowable compressive force in reinforced masonry due to axial load	kips	N
P_e	Euler buckling load	kips	N
P_{nw}	nominal axial strength of a wall	kips	N
P_o	maximum allowable axial load for zero eccentricity	kips	N
q	allowable horizontal shear per shear connector	kips	N
q	soil pressure under footing	lbf/ft^2	Pa
q	uniform surcharge	lbf/ft, lbf/ft^2	N/m, Pa
q_c	tip resistance	lbf/ft^2	Pa
q_s	skin friction resistance	lbf/ft^2	Pa
q_z	velocity pressure	lbf/ft^2	Pa
Q	bearing capacity	lbf/ft^2	Pa
Q	nominal load effect	various	various
Q	statical moment	in^3	mm^3
r	radius of gyration	in, ft	mm, m
r	rigidity	–	–
R	allowable load per bolt	kips	N
R	concentrated load	lbf, kips	N
R	reaction	lbf, kips	N
R	resultant force	lbf	N
RF	reduction factor	–	–
s	spacing	in	mm
S	force	lbf	N
S	section modulus	in^3, ft^3	mm^3, m^3
S	snow load	kips	N

S_{DS}	design earthquake spectral response accelerations at short period	–	–
S_{MS}	maximum considered earthquake spectral response accelerations at short period	–	–
t	nominal weld size	in	mm
t	slab thickness	in	mm
t	thickness	in, ft	mm, m
t	wall thickness	in, ft	mm, m
t_e	effective throat thickness of a weld	in	mm
T	period	sec	s
T	temperature	°F, °R	°C, K
T	tension force	lbf, kips	N
T	torsional moment (torque)	ft-lbf	N·m
T_u	factored torsional moment	ft-lbf	N·m
u	unit force	–	–
U	ultimate strength required to resist factored loads	lbf	N
v	wind speed	mph	kph
v	shear stress	lbf/in^2	Pa
v_c	allowable concrete shear stress	lbf/in^2	Pa
v_u	shear stress due to factored loads	lbf/in^2	Pa
V	design shear force	lbf	N
V	shear	lbf, kips	N
V	shear strength	lbf, lbf/in^2	N, Pa
V_c	allowable concrete shear strength	lbf	N
V_u	factored shear force	lbf	N
w	distributed load	lbf/ft	N/m
w	uniformly distributed load	lbf/ft, lbf/ft^2	N/m, Pa
w	weld size	in	mm
w'	tributary width	in, ft	mm, m
W	effective seismic weight	lbf, kips	N
W	nail withdrawal value	lbf	N
W	weight	lbf	N
W	wind load	lbf	N
x	distance	in, ft	mm, m
x	location	in	mm
x	x-coordinate of position	ft	m
$\bar{x}$	distance to center of rigidity in the x-direction	in, ft	mm, m
$\bar{y}$	distance from centroidal axis to the centroid of the area	in	mm
$\bar{y}$	distance to center of rigidity in the y-direction	in, ft	mm, m
y	location	in	m
y	y-coordinate of position	ft	m
y_c	distance from top of the section to the centroid of the section	in	mm
y_t	distance from centroid of the section to the extreme fiber in tension	in	mm
Z	connector lateral design value	lbf	N

Symbols

α	coefficient of linear thermal expansion	°F^{-1}	°C^{-1}
α	ratio of flexural stiffness of beams in comparison to slab	–	–
β	column strength factor	–	–
β	ratio of clear spans in long-to-short directions of a two-way slab	–	–
β	ratio of long side to short side of a footing	–	–
β_{dns}	ratio accounting for column stiffness reduction due to sustained axial load	–	–
γ	ratio of the distance between bars on opposite faces of a column to the overall column dimension, both measured in the direction of bending	–	–
γ	specific weight (unit weight)	lbf/ft^3	N/m^3
Δ	deflection	in	mm
θ	angle	deg, rad	deg, rad
λ	distance from centroid of compressed area to extreme compression fiber	in	mm
λ	height and exposure adjustment factor	–	–
ρ	reinforcement ratio	–	–
ρ_g	longitudinal reinforcement ratio	–	–
ρ_h	ratio of horizontal shear reinforcement to gross concrete area	–	–
ρ_n	ratio of vertical shear reinforcement area to gross concrete area for a shear wall	–	–
σ	normal stress	lbf/ft^2	Pa
τ	shear stress	lbf/ft^2	Pa
ϕ	strength reduction factor	–	–
Ψ	relative stiffness parameter	–	–

Subscripts

γ	density
0	initial
a	active, allowable, or due to axial loading
A	adhesion
b	beam, bending, bolt, or masonry breakout
bm	beam
bot	bottom
BS	block shear
c	cohesive, column, compression, concrete, curvature, or masonry crushing
cr	cracked or cracking
cs	critical section
ct	condition treatment
d	directionality or penetration depth
D	dead load or duration
e	effective or exposure
eff	effective
eg	end grain
E	Euler
f	flange, flat, force, form, or skin friction
fs	face shell
ftg	footing
fu	flat use
F	size
g	gross or grout
h	horizontal
i	initial, inside, or ith member
j	jth member
k	kern or kth member
l	longitudinal
ls	load sharing
L	beam stability, left, or live load
m	masonry
max	maximum
min	minimum
M	wet service
n	nail, net, or nominal
nom	nominal
ns	nonsway frame
o	centroidal or initial
OT	overturning
p	bearing, column stability, passive, pile tip, plate, prestressed, or projecting
prov	provided
pry	anchor pryout
q	surcharge
r	rafter, reduced, repetitive, resultant, rib, roof, or rupture
R	resistance, resisting, resistive, resultant, or right
req	required
s	shear, side friction, simplified, skin, snow, spiral, steel reinforcement, or steel yieldin
sat	saturated
sd	short direction
st	steel
SR	stress range
SL	sliding
t	temperature, tensile, tension, topography, torsion, or tributary
th	thermal
tr	transformed
u	ultimate (factored), ultimate tensile, or unbraced
v	shear or vertical
V	volume
w	wall, web, weld, or wind
x	at a distance x, in x-direction, or strong axis
y	in y-direction, weak axis, or yield
z	at height z

1 Analysis of Structures

FORCES AND LOAD EFFECTS

PROBLEM 1

A concrete column and footing carries the loads shown. The footing is 6 ft wide.

column dead load	80 kips
column live load	100 kips
moment due to wind	240 ft-kips
compressive strength of concrete	3000 lbf/in^2
maximum allowable soil pressure	4000 lbf/ft^2

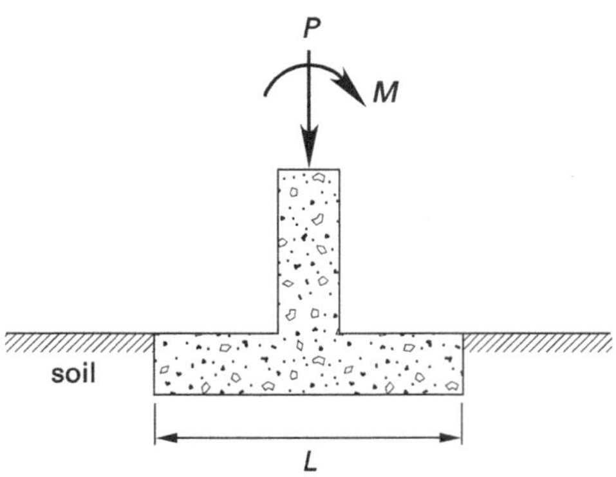

What is the minimum footing length, L, required for the entire footing to be considered effective in carrying these loads?

(A) 7.7 ft

(B) 8.0 ft

(C) 8.7 ft

(D) 10 ft

Hint: For the entire footing to be effective, avoid tensile stress in the soil.

PROBLEM 2

The cantilevered beam shown has a varying moment of inertia. The moment of inertia for the first 15 ft, I_1, is 2000 in^4. The second moment of inertia, I_2, is 1000 in^4. The modulus of elasticity of the beam is 29×10^6 lbf/in^2.

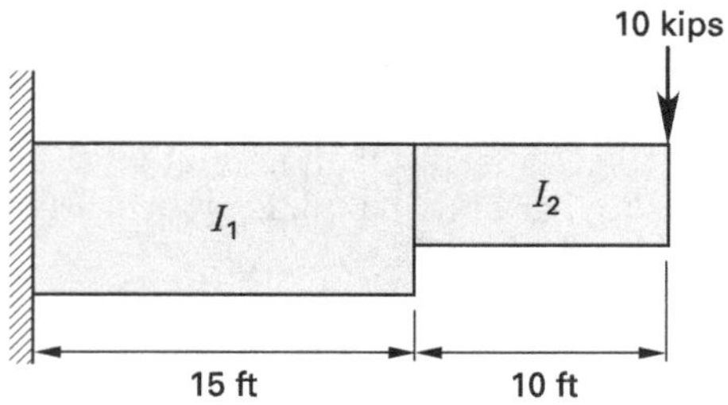

The deflection at the free end of the beam is most nearly

(A) 0.0017 in

(B) 1.7 in

(C) 3.0 in

(D) 3.1 in

Hint: Use the moment-area method to find the deflection of the beam.

PROBLEM 3

A column with the cross section shown carries an axial load of 560 lbf applied at the centroid of the section.

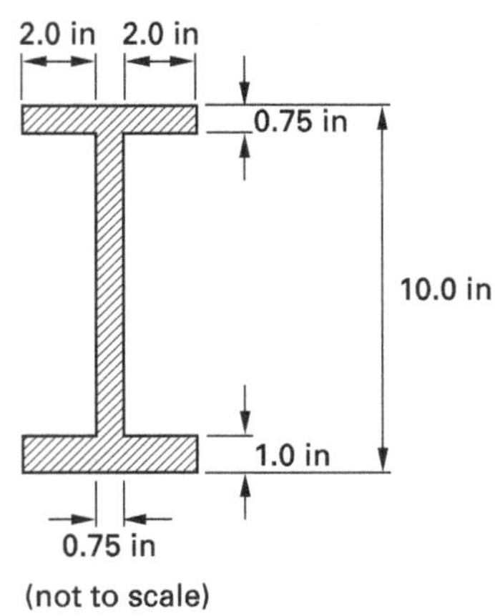

The compressive stress for the cross section is most nearly

(A) 15 lbf/in^2

(B) 35 lbf/in^2

(C) 39 lbf/in^2

(D) 42 lbf/in^2

Hint: An axial load applied at the centroid of the section creates uniformly distributed stress.

PROBLEM 4

A 2.5 in diameter mild steel bar is fixed at each end by a steel plate. The modulus of elasticity of the steel is 29×10^6 lbf/in^2. At 8:00 a.m., the bar has a temperature of 35°F and experiences zero stress. At 3:30 p.m., the temperature of the bar is 100°F. Ignore gravity loads. The axial load in the bar at 3:30 p.m. is most nearly

(A) 60 kips

(B) 92 kips

(C) 240 kips

(D) 6000 kips

Hint: Chapter 2 of the AISC *Steel Construction Manual* contains information on section properties and thermal expansion properties of steel.

PROBLEM 5

A one-story steel-framed building has earth bermed on one side. The total resultant load from the earth pressure is 600 kips and is resisted entirely by the three shear walls as shown. Wall A is 12 in thick. Walls B and C are 14 in thick. The roof is framed with steel joists and corrugated steel deck (no concrete).

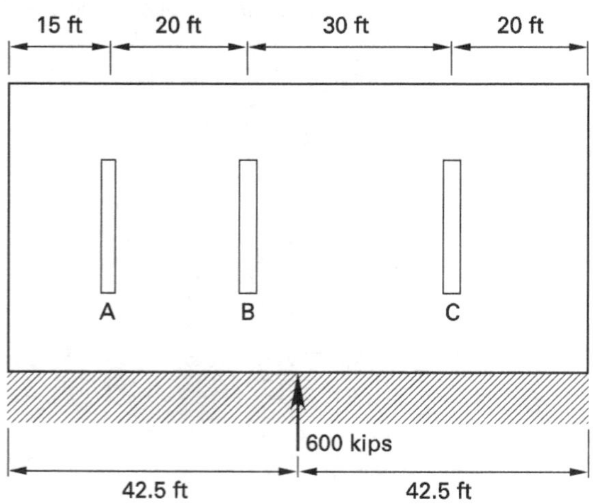

What is most nearly the shear load on wall A?

(A) 124 kips

(B) 176 kips

(C) 180 kips

(D) 247 kips

Hint: A steel joist and corrugated deck roof system is typically considered a flexible diaphragm unless a concrete slab is poured on the roof deck.

PROBLEM 6

A beam with a constant cross section and constant modulus of elasticity is uniformly loaded as shown. The relative stiffness of section AB is 0.286/ft and of section BC is 0.190/ft.

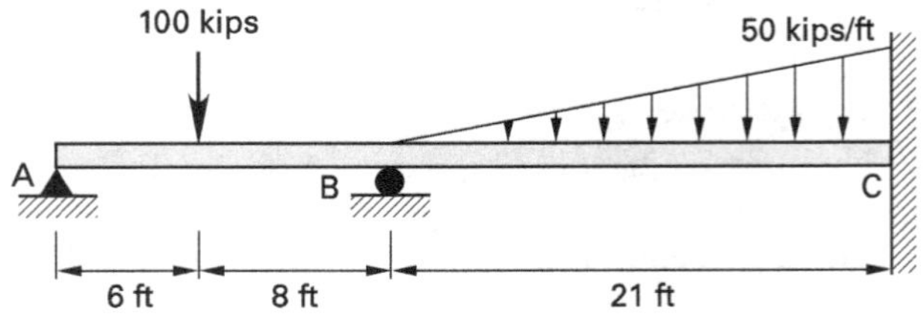

Using moment distribution, the unbalanced portion of the fixed-end moments at joint B is most nearly

(A) 490 ft-kips, clockwise

(B) 490 ft-kips, counterclockwise

(C) 540 ft-kips, clockwise

(D) 960 ft-kips, clockwise

Hint: Consult a reference containing fixed-end moments.

PROBLEM 7

For the one-way beam shown, use the moment coefficients found in ACI 318 to determine the largest magnitude moment and its location. Support A is a beam, supports BC and DE are built-in columns, and support F is a simple support. The beam carries a uniform factored load of 600 lbf/ft, and the live load is 1.5 times the unfactored dead load.

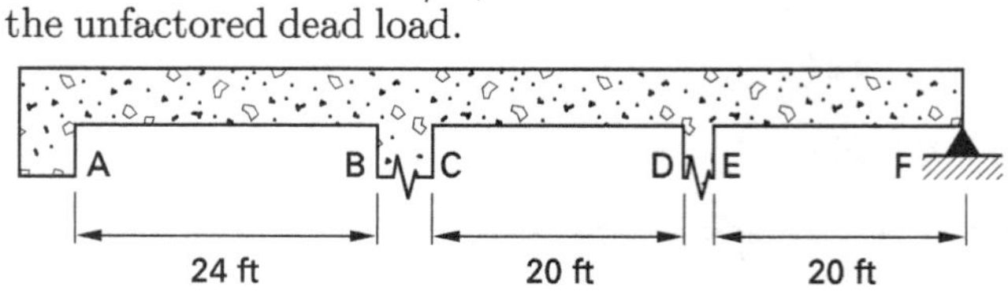

The moment at point B is most nearly

(A) −38,000 ft-lbf

(B) −35,000 ft-lbf

(C) −29,000 ft-lbf

(D) +25,000 ft-lbf

Hint: Refer to ACI 318 Sec. 6.5.

PROBLEM 8

The 10 ft long channel beam shown is made from ½ in thick flat plates welded together and is simply supported at both ends. A load is applied at the midspan of the beam.

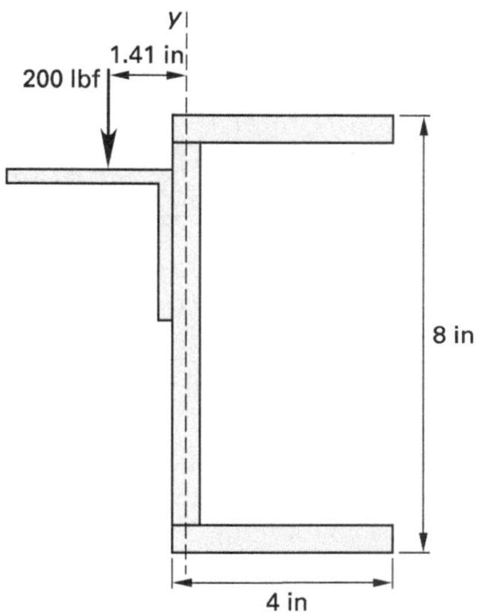

The load on the beam is most nearly

(A) 280 in-lbf (torsion)

(B) 280 ft-lbf (torsion)

(C) 280 in-lbf (torsion), 500 ft-lbf (flexure)

(D) 500 ft-lbf (flexure)

Hint: Determine the shear center of the channel.

PROBLEM 9

Analyze the truss shown.

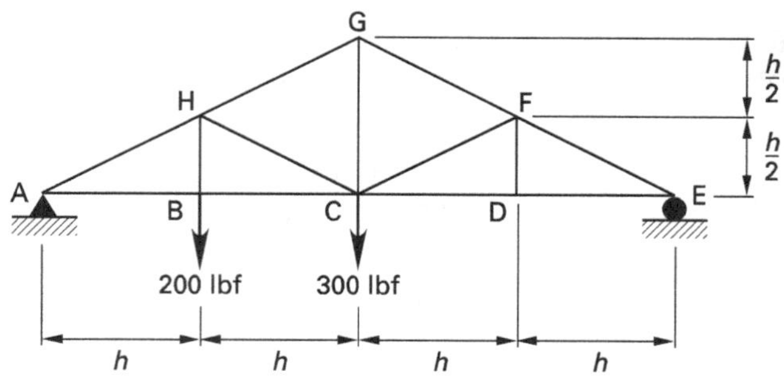

The force in member AH is most nearly

(A) 300 lbf (compression)

(B) 600 lbf (compression)

(C) 670 lbf (compression)

(D) 670 lbf (tension)

Hint: Solve by using the method of joints.

PROBLEM 10

A 20 ft simply supported beam carries a load that decreases linearly from a maximum of 350 lbf/ft at its left support to 0 at the midspan of the beam, as shown.

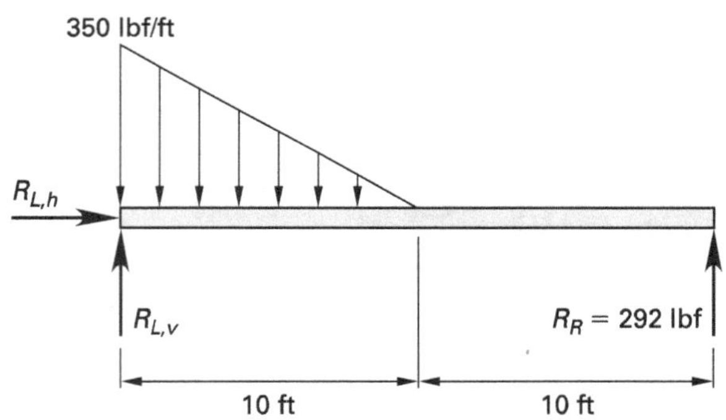

If R_R is equal to 292 lbf, the point at which the shear is zero measured from the left support is most nearly

(A) 0.87 ft

(B) 4.1 ft

(C) 5.9 ft

(D) 14 ft

Hint: Write the equation for shear measured from the right support.

PROBLEM 11

A three-span continuous steel beam carries a uniform dead load of 500 lbf/ft and a uniform live load of 1000 lbf/ft on all three spans. It is an ASTM A992 W16 × 31 beam, and each span is 20 ft.

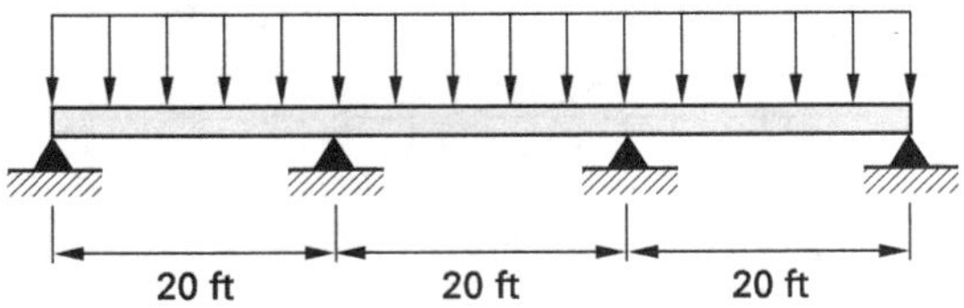

What is the maximum deflection of the first span?

(A) 0.21 in

(B) 0.27 in

(C) 0.33 in

(D) 0.51 in

Hint: Use the beam loading diagrams found in the AISC *Steel Construction Manual*.

PROBLEM 12

The T-shaped beam shown carries a 100 lbf shear load.

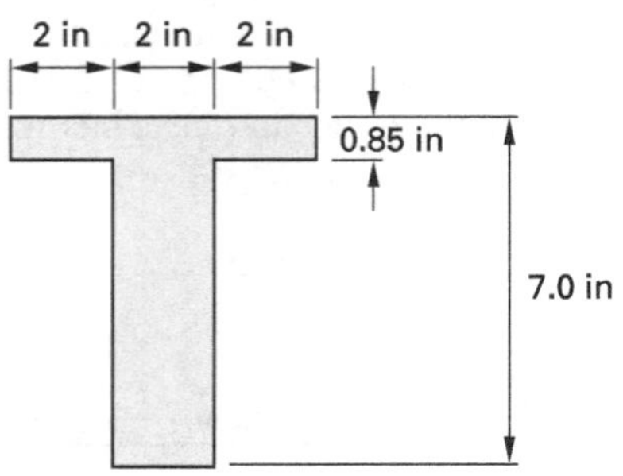

The shear stress at the centroid of the cross section is most nearly

(A) 6.5 lbf/in^2

(B) 8.1 lbf/in^2

(C) 10 lbf/in^2

(D) 22 lbf/in^2

Hint: First, find the centroid of the cross section.

PROBLEM 13

A building is constructed with 9 ft high shear walls as shown.

wall A: 16 in thick, 60 ft long
wall B: 16 in thick, 36 ft long
wall C: 12 in thick, 40 ft long
wall D: 12 in thick, 36 ft long
wall E: 16 in thick, 36 ft long

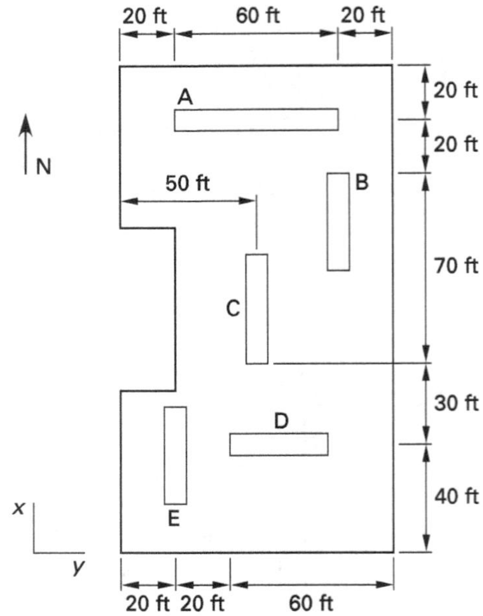

The center of rigidity of the building is located at a point that is most nearly

(A) 115 ft from the south edge of the building and 50.0 ft from the east edge of the building

(B) 115 ft from the south edge of the building and 54.9 ft from the west edge of the building

(C) 123 ft from the north edge of the building and 50.0 ft from the east edge of the building

(D) 123 ft from the south edge of the building and 50.0 ft from the west edge of the building

Hint: The center of rigidity can be based on the relative areas of the walls.

PROBLEM 14

A building is designed with walls that act in shear only to resist the wind load as shown. The north wind load is 200 lbf/ft. All of the walls are 12 in thick. The center of rigidity (c.r.) of the building is located 98.3 ft from the west facade and 40 ft from the south facade.

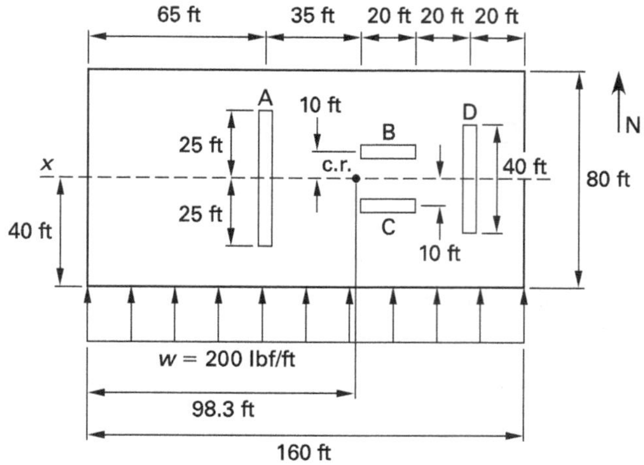

What is most nearly the shear load on wall A?

(A) 6.6 kips

(B) 12 kips

(C) 18 kips

(D) 25 kips

Hint: Wind load is distributed to shear walls according to the relative rigidity of the walls.

PROBLEM 15

A pin-connected truss is used to support a sign as shown. The product of the area and the modulus of elasticity equals 3 kips for all members except AD and BC, for which the product of the area and the modulus of elasticity equals 5 kips.

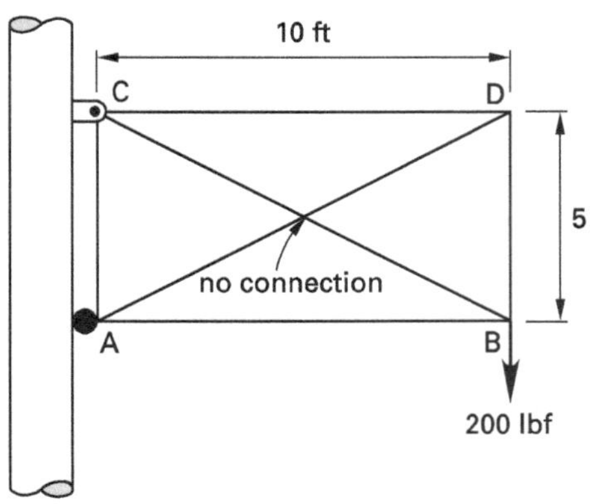

What is most nearly the force in member BC?

(A) −210 lbf (compression)

(B) 230 lbf (tension)

(C) 240 lbf (tension)

(D) 450 lbf (tension)

Hint: Assess whether the truss is determinate.

PROBLEM 16

A moment distribution analysis of a beam determined the moments at the beam supports as shown.

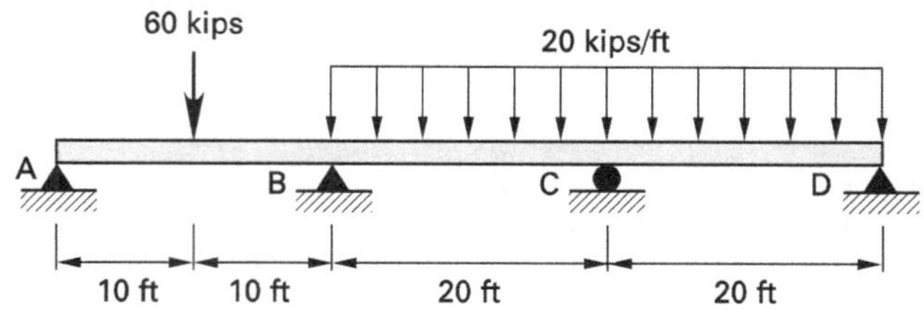

	A	B		C		D
DF	1.0	0.5	0.5	0.5	0.5	1.0
FEM	−150.0	+150.0	−666.7	+666.7	−666.7	+666.7

Moment distribution gives the following results.

	A	B		C		D
M	0	+386.8	−386.8	+903.2	−903.2	0

The reaction at support B is most nearly

(A) 50 kips

(B) 120 kips

(C) 170 kips

(D) 220 kips

Hint: Use free-body diagrams to determine the shear at the supports.

PROBLEM 17

A flat-plate concrete slab is supported by 12 in square concrete columns. The critical factored load on the slab is 168 lbf/ft². The design moments of a typical interior slab panel shown are to be determined using the direct design method.

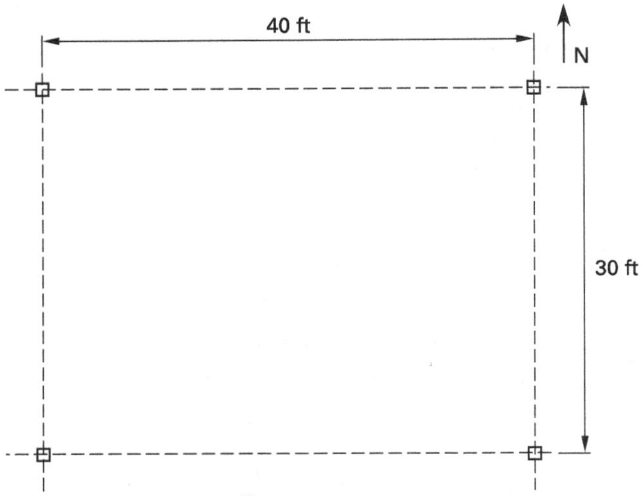

The midspan moment of the middle strip in the east/west direction is most nearly

(A) 77 ft-kips

(B) 99 ft-kips

(C) 130 ft-kips

(D) 140 ft-kips

Hint: Refer to ACI 318 Chap. 8.

PROBLEM 18

The beam shown has a modulus of elasticity of 1.2×10^6 lbf/in^2 and a moment of inertia of 240 in^4.

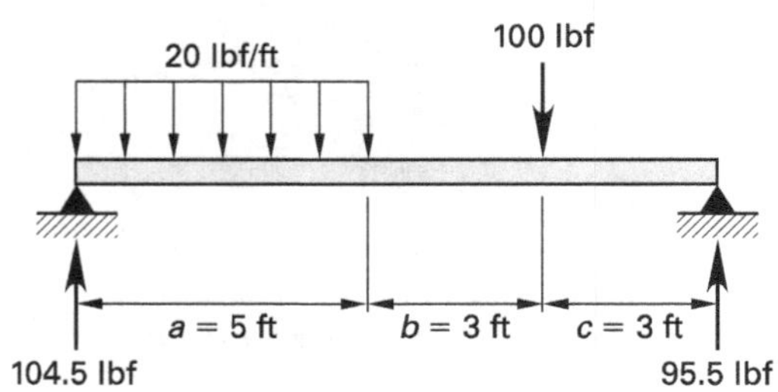

The deflection at a point 5 ft from the left support of the beam is most nearly

(A) 0.0099 in

(B) 0.011 in

(C) 0.013 in

(D) 0.022 in

Hint: Use AISC *Steel Construction Manual* Table 3-23.

PROBLEM 19

The beam shown carries a uniform eccentric load of 20 kips/ft. The span is 20 ft, and the beam is fully restrained at one end.

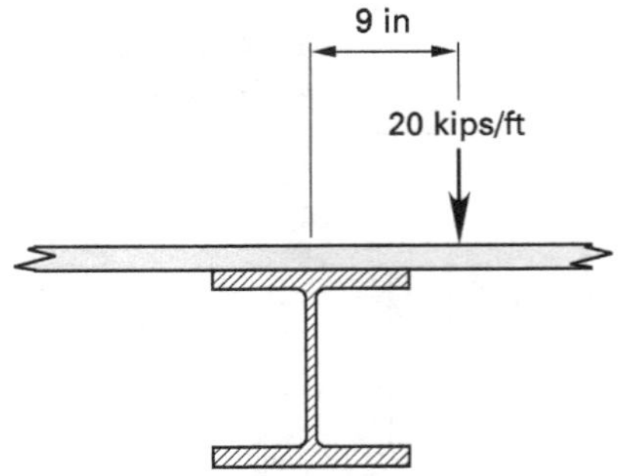

What is most nearly the maximum torsional moment on the beam?

(A) 150 ft-kips

(B) 300 ft-kips

(C) 670 ft-kips

(D) 1000 ft-kips

Hint: Torsion is the product of a force and an eccentricity.

LOADS AND LOAD APPLICATIONS

PROBLEM 20

The three-story brick veneer office building shown is framed with structural steel. Assume simple connections, full lateral support of the beams, and uniform dead and live loads. The floor-to-floor height of each story is 13 ft.

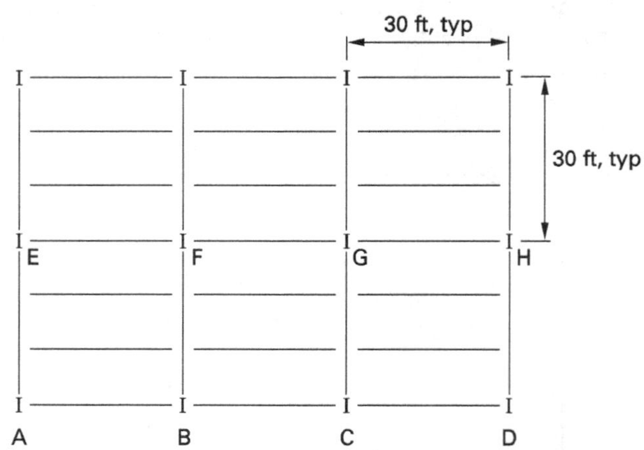

partial framing plan

Which of the following statements is true?

(A) The total load on spandrel beam BC is equal to half of the total load on interior beam FG.

(B) The total load on spandrel beam BC is equal to one-third of the total load on spandrel beam AE.

(C) The live load on spandrel beam AE is equal to twice the live load on spandrel beam AB.

(D) The total load on beam AE is equal to the total load on beam AB.

Hint: The weight of the facade is typically supported by spandrel beams at each floor.

PROBLEM 21

A 100 acre office park is located in a flat, suburban area in South Dakota. All buildings in the office park are 45 ft high and have a 40 ft by 60 ft footprint. The office park's landscaping uses only low-level ground cover. According to the *International Building Code* (IBC), a building in the office park would have a snow exposure factor of

(A) 0.8

(B) 0.9

(C) 1.0

(D) 1.2

Hint: The IBC refers to ASCE/SEI7 Chap. 7 for determining a building's snow exposure factor.

PROBLEM 22

A continuous L4 × 6 × ³⁄₈ angle anchored to a 12 in concrete masonry wall is used as a ledger for floor joists as shown. A307 headed anchor bolts are placed 16 in on center in grouted cells. The joists are spaced 6 ft on center. The specified compressive strength of masonry is 1500 lbf/in^2.

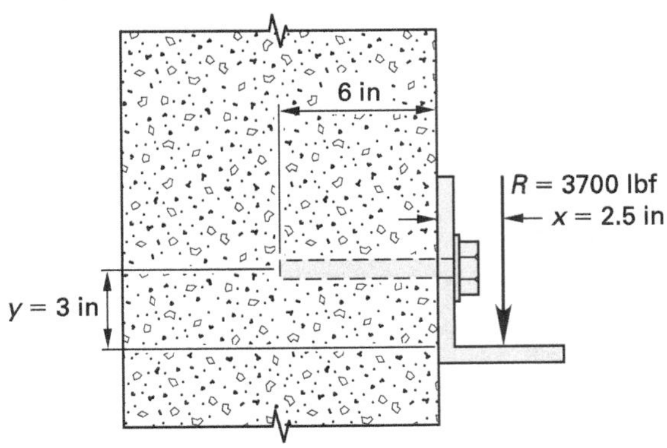

If the load from the joists is 3700 lbf, the load on each anchor bolt is most nearly

(A) 460 lbf shear, 390 lbf tension

(B) 820 lbf shear, 0 lbf tension

(C) 820 lbf shear, 690 lbf tension

(D) 3700 lbf shear, 3100 lbf tension

Hint: Anchor bolts are subject to both shear and tension.

PROBLEM 23

A two-story office building has a 60 ft × 100 ft rectangular floor plan. Columns spaced 20 ft apart carry the following loads from the roof and second floor. Live load reductions are not permitted.

roof dead load	15 lbf/ft^2
roof live load	20 lbf/ft^2
floor dead load	15 lbf/ft^2
floor live load	60 lbf/ft^2

What is most nearly the total load on an interior first-floor column using the *International Building Code* (IBC) basic load combinations for allowable stress design?

(A) 1.8 kips

(B) 14 kips

(C) 36 kips

(D) 44 kips

Hint: Determine the tributary area for an interior column.

PROBLEM 24

A steel beam supports a 6 in (nominal) lightweight concrete masonry (CMU) wall around a mechanical room. The CMU has a unit weight, γ, of 85 lbf/ft^3, and the wall is ungrouted. If the wall is 8 ft high and 10 ft long, the load on the beam is most nearly

(A) 20 lbf/ft

(B) 160 lbf/ft

(C) 200 lbf/ft

(D) 320 lbf/ft

Hint: The load on the beam is uniformly distributed.

PROBLEM 25

A 20 ft tall, single-story flex warehouse building is constructed of 12 in normalweight concrete masonry walls laid in running bond that have been grouted solid. One 20 ft long wall contains a 4 ft opening centered in the wall as shown. An 8 in deep concrete masonry bond beam spans the opening. A 700 lbf concentrated load from a roof truss is centered over the opening.

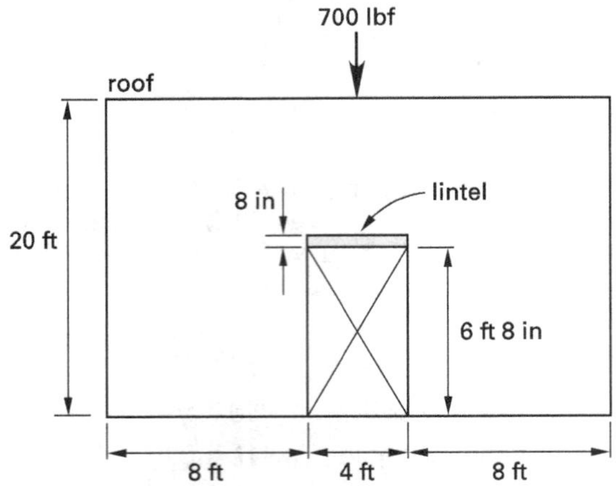

The design moment on the lintel includes

(A) only the weight of the wall directly above the door opening

(B) the roof load plus the weight of the wall directly above the door opening

(C) only the weight of the wall within a 45° triangle over the door opening

(D) the roof load plus the weight of the wall within a 45° triangle over the door opening

Hint: The wall described will exhibit arching action over the opening.

PROBLEM 26

The three-story, 4 in brick veneer (40 lbf/ft^2) office building shown is framed with ASTM A992 structural steel. Assume simple connections and full lateral support. The floor dead load is 60 lbf/ft^2, and the floor live load is 40 lbf/ft^2. The beam's self-weight is 50 lbf/ft.

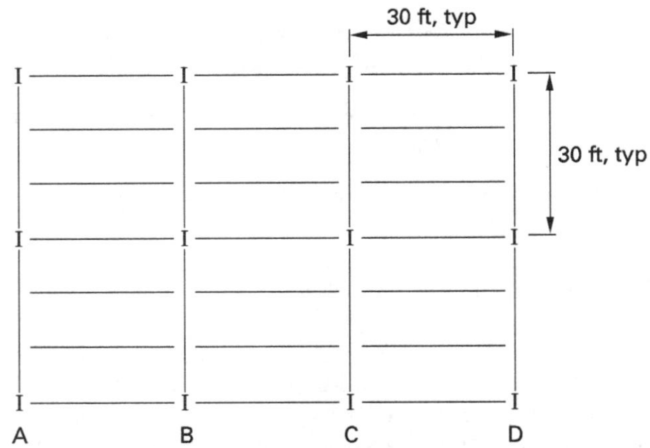

partial framing plan

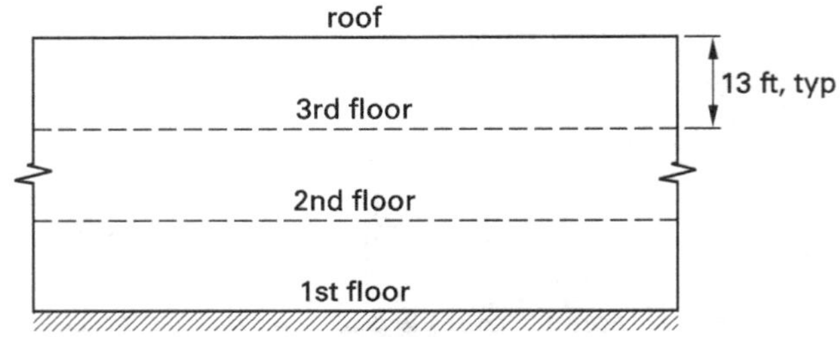

partial facade elevation

The maximum factored design moment in 2nd floor spandrel beam BC is most nearly

(A) 83.0 ft-kips

(B) 120 ft-kips

(C) 137 ft-kips

(D) 153 ft-kips

Hint: Include the weight of the facade in the design moment calculation.

SOLUTION 1

Limit eccentricity, e, to one-sixth of the footing length to keep the soil resultant within the kern limit and avoid tension in the soil.

$$e = \frac{M}{P} = \frac{M_{\text{wind}}}{P_D + P_L} = \frac{240 \text{ ft-kips}}{80 \text{ kips} + 100 \text{ kips}}$$
$$= 1.33 \text{ ft} \quad \left[\leq \tfrac{1}{6}L\right]$$

$$e \leq \frac{L}{6}$$
$$1.33 \text{ ft} \leq \frac{L}{6}$$

Solving for L,

$$e = \tfrac{1}{6}L$$
$$L = 6e$$
$$= (6)(1.33 \text{ ft})$$
$$= 7.98 \text{ ft} \quad (8.0 \text{ ft})$$

To complete the footing design, the allowable soil pressure would be checked and the footing size adjusted accordingly.

The answer is (B).

Why Other Options Are Wrong

(A) This incorrect solution calculates the required size of a square footing (ignoring the footing width given) based on the dead and live loads and allowable soil pressure. The problem statement asks for the minimum length required for the entire footing to be effective, not the minimum size based on the soil pressure.

(C) This incorrect solution calculates eccentricity using factored loads. Factored loads are used in concrete design but not for determining footing size. Footing size should be based on unfactored soil pressure.

(D) This incorrect solution calculates the required length based on the dead and live loads and allowable soil pressure. The problem statement asks for the minimum length required for the entire footing to be effective, not the minimum size based on the soil pressure.

SOLUTION 2

Using the moment-area method, the deflection of a beam at a particular point is equal to the moment of the M/EI diagram about that point.

The moment at the fixed end of the beam shown is

$$M = Pl = (10 \text{ kips})(25 \text{ ft})$$
$$= 250 \text{ ft-kips}$$

The moment diagram is

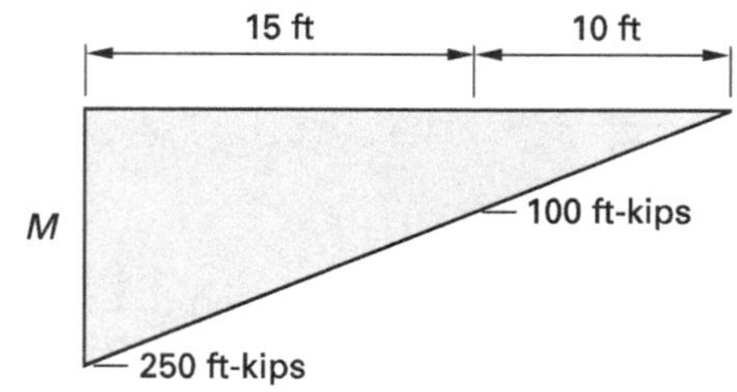

To draw the M/EI diagram, divide the moment diagram by the respective moment of inertia.

$$\frac{M_{\text{support}}}{EI} = \frac{250 \text{ ft-kips}}{E(2000 \text{ in}^4)} = \frac{0.125 \text{ ft-kips}}{E \text{ in}^4}$$

The M/EI diagram is

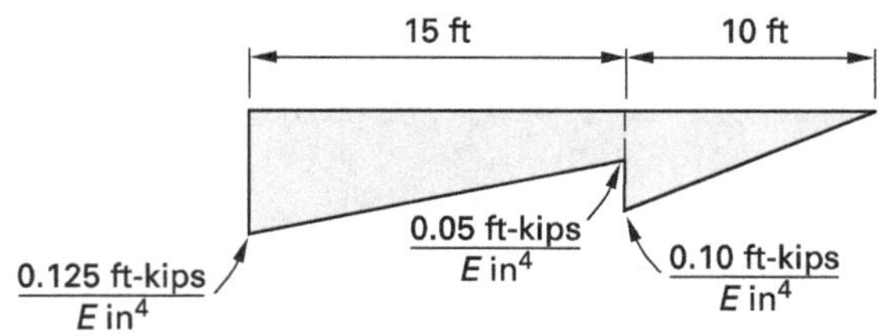

The deflection at the free end of the beam is the moment of the M/EI diagram about the free end.

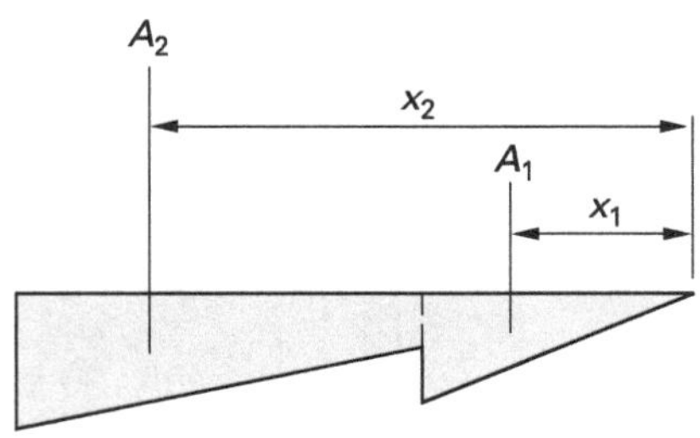

$$\begin{aligned}\delta &= A_1x_1 + A_2x_2\\ &= \left(\frac{1}{2}\right)(10\text{ ft})\left(\frac{0.10\text{ ft-kips}}{E\text{ in}^4}\right)\left(\frac{2}{3}\right)(10\text{ ft})\\ &\quad +\left(\frac{0.05\text{ ft-kips}}{E\text{ in}^4}\right)(15\text{ ft})\left(10\text{ ft}+\frac{15\text{ ft}}{2}\right)\\ &\quad +\left(\frac{1}{2}\right)(15\text{ ft})\left(\frac{0.125\text{ ft-kips}-0.05\text{ ft-kips}}{E\text{ in}^4}\right)\\ &\quad \times\left((15\text{ ft})\left(\frac{2}{3}\right)+10\text{ ft}\right)\\ &= \frac{27.7\ \dfrac{\text{ft}^3\text{-kips}}{\text{in}^4}}{E}\\ &= \left(\frac{27.7\ \dfrac{\text{ft}^3\text{-kips}}{\text{in}^4}}{29\times10^6\ \dfrac{\text{lbf}}{\text{in}^2}}\right)\left(12\ \frac{\text{in}}{\text{ft}}\right)^3\left(1000\ \frac{\text{lbf}}{\text{kip}}\right)\\ &= 1.65\text{ in}\quad(1.7\text{ in})\end{aligned}$$

The answer is (B).

Why Other Options Are Wrong

(A) This incorrect solution does not convert kips to pounds in calculating the deflection. The units do not work out.

(C) This incorrect solution reverses I_1 and I_2 when drawing the M/EI diagram.

(D) This incorrect solution does not make the adjustment for the differing moment of inertia over the span length. The deflection is incorrectly calculated using a moment of inertia of 1000 in^4 for the entire length of the beam.

SOLUTION 3

Find the total area of the section.

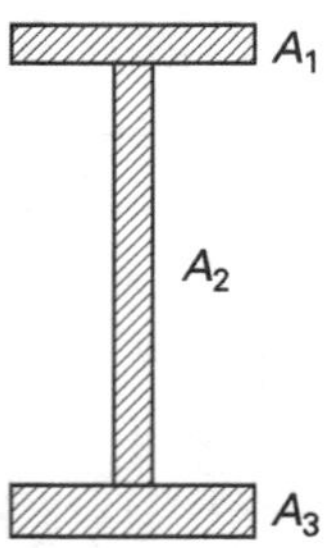

$$\begin{aligned}A &= A_1 + A_2 + A_3\\ &= (0.75\text{ in})(4.75\text{ in}) + (10.0\text{ in} - 0.75\text{ in} - 1.0\text{ in})\\ &\qquad \times(0.75\text{ in}) + (1.0\text{ in})(4.75\text{ in})\\ &= 14.5\text{ in}^2\end{aligned}$$

The compressive stress is

$$\sigma = \frac{P}{A} = \frac{560\text{ lbf}}{14.5\text{ in}^2} = 38.6\text{ lbf/in}^2\quad(39\text{ lbf/in}^2)$$

The answer is (C).

Why Other Options Are Wrong

(A) This incorrect solution calculates the section area, not the stress.

(B) This incorrect solution miscalculates area A_2 by using the overall length of the section (10 in) for the length of A_2.

(D) This incorrect solution miscalculates A_1 and A_3 by using 4.0 in for the width of both sections instead of 4.75 in.

SOLUTION 4

The induced axial load in a constrained member due to a temperature change is

$$P_{\text{th}} = \alpha(T_2 - T_1)AE$$

α is the coefficient of thermal expansion of the member. According to Chap. 2 of the AISC *Steel Construction Manual*, the coefficient of thermal expansion for mild steel at temperatures up to 100°F is

$$\alpha = \frac{0.00065}{100^\circ\text{F}}$$

The area of the steel bar is

$$A = \pi\left(\frac{d^2}{4}\right) = \pi\left(\frac{(2.5\text{ in})^2}{4}\right) = 4.91\text{ in}^2$$

$$\begin{aligned}P_{\text{th}} &= \alpha(T_2 - T_1)AE\\ &= \left(\frac{0.00065}{100^\circ\text{F}}\right)(100^\circ\text{F} - 35^\circ\text{F})(4.91\text{ in}^2)\\ &\qquad\times\left(29\times10^6\ \frac{\text{lbf}}{\text{in}^2}\right)\left(\frac{1\text{ kip}}{1000\text{ lbf}}\right)\\ &= 60.2\text{ kips}\quad(60\text{ kips})\end{aligned}$$

The answer is (A).

Why Other Options Are Wrong

(B) This incorrect solution uses the maximum temperature instead of the temperature change to calculate the induced axial load.

(C) This solution incorrectly calculates the area of the bar as πd^2 instead of $\pi d^2/4$.

(D) This incorrect solution reads the coefficient of thermal expansion as having no units instead of as 0.00065 per 100°F. The units do not work out in the axial load equation.

SOLUTION 5

A roof framed with steel joists and corrugated steel deck is typically considered a flexible diaphragm because of its light weight and flexibility. A rigid diaphragm can be achieved if 2–3 in of concrete are used on the steel decking.

For a flexible diaphragm, the lateral loads are distributed according to tributary area. If the roof were a rigid diaphragm (i.e., concrete), the lateral loads would be distributed according to rigidity.

In this case, the shear load can be distributed according to the tributary width, w.

$$V_{\text{A}} = \left(\frac{w_{\text{A}}}{\sum w_i}\right) V_{\text{total}} = \left(\frac{15 \text{ ft} + \left(\frac{1}{2}\right)(20 \text{ ft})}{85 \text{ ft}}\right)(600 \text{ kips})$$
$$= 176 \text{ kips}$$

The answer is (B).

Why Other Options Are Wrong

(A) This incorrect solution calculates the tributary width of wall A using one-half the distance to the edge of the building instead of the full tributary width.

(C) This incorrect solution distributes the lateral load based on relative rigidities. Relative rigidities can only be used if the diaphragm is rigid. In addition, when loads are based on rigidities, eccentricity of the load (torsion) must also be considered.

(D) This incorrect solution calculates the tributary width of wall A using the full distance between walls instead of one-half the distance between the walls.

SOLUTION 6

The fixed-end moments (FEM) at joint B, as taken from a reference text, are

$$\begin{aligned} \text{FEM}_{\text{BA}} &= \frac{P}{L^2}\left(b^2 a + \frac{a^2 b}{2}\right) \\ &= \left(\frac{100 \text{ kips}}{(14 \text{ ft})^2}\right)\left((6 \text{ ft})^2(8 \text{ ft}) + \frac{(8 \text{ ft})^2(6 \text{ ft})}{2}\right) \\ &= 244.9 \text{ ft-kips} \end{aligned}$$

$$\text{FEM}_{\text{BC}} = \frac{wL^2}{30} = \frac{\left(50 \frac{\text{kips}}{\text{ft}}\right)(21 \text{ ft})^2}{30} = 735.0 \text{ ft-kips}$$

The unbalanced moment at joint B is

$$\begin{aligned} &\text{FEM}_{\text{BA}} - \text{FEM}_{\text{BC}} \\ &\quad = 244.9 \text{ ft-kips} - 735.0 \text{ ft-kips} \\ &\quad = -490.1 \text{ ft-kips} \quad (490 \text{ ft-kips, clockwise}) \end{aligned}$$

The answer is (A).

Why Other Options Are Wrong

(B) This incorrect solution determines the unbalanced moment correctly as in the solution. However, the sign convention is not applied correctly, and a counterclockwise rotation is assumed.

(C) This incorrect solution reverses the distribution factors at joint B, putting the distribution factor for BA at BC and vice versa.

(D) This incorrect solution reverses the FEMs for the triangular load.

SOLUTION 7

Section 6.5 of ACI 318 gives the moment at a given point along a one-way beam based on the moment coefficient, C_1.

$$M = C_1 wL^2$$

Moment coefficients can only be used under the following conditions: there are two or more spans; spans are approximately equal, with the larger of two adjacent spans not greater than the shorter by more than 20%; loads are uniformly distributed; the unfactored live load does not exceed three times the unfactored dead load; or members are prismatic.

The larger span (24 ft) is 20% greater than the smaller span (20 ft), so moment coefficients can be used.

The clear span, l_n, is defined in ACI Chap. 2 and Table 6.5.2 as equal to the clear span for positive moments or shear, and as the average of adjacent clear spans for negative moments.

The moment at point B (the exterior face of the first interior support) is

$$\begin{aligned} M_{\text{B}} &= -\frac{1}{10} w_u l_n^2 = \left(-\frac{1}{10}\right)\left(600 \frac{\text{lbf}}{\text{ft}}\right)\left(\frac{24 \text{ ft} + 20 \text{ ft}}{2}\right)^2 \\ &= -29{,}040 \text{ ft-lbf} \quad (-29{,}000 \text{ ft-lbf}) \end{aligned}$$

The answer is (C).

Why Other Options Are Wrong

(A) This incorrect solution uses the moment coefficients for a two-span condition instead of those for a three-span condition and a value of l_n equal to 24 ft instead of $\frac{1}{2}$(24 ft + 20 ft), or 22 ft.

(B) This incorrect solution does not use the average of adjacent clear spans when determining the negative moment at point B.

(D) This incorrect solution calculates the moment for span AB.

SOLUTION 8

Beams with eccentric loads are subject to torsion if the center of the load does not pass through the shear center of the shape. For symmetrical shapes (such as W shapes), the shear center is at the centroid of the shape. From the *Structural Engineering Reference Manual* or another reference, for the channel beam shown, the shear center is located at a distance from the midpoint of the vertical leg of $e = 3b^2/h + 6b$.

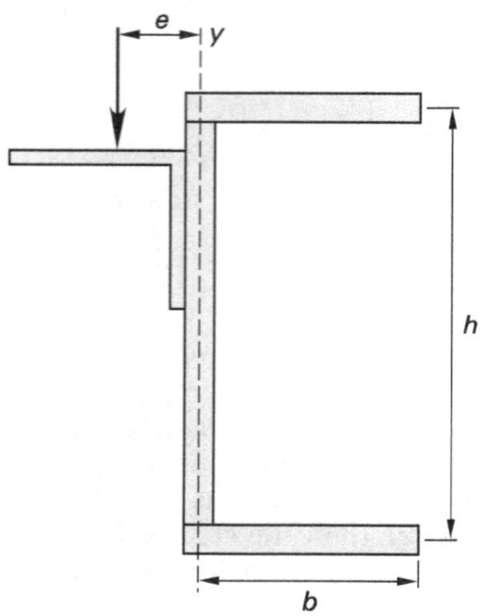

In this case,

$$b = 4.0\ \text{in} - \frac{t}{2} = 4.0\ \text{in} - \frac{0.5\ \text{in}}{2} = 3.75\ \text{in}$$
$$h = 8.0\ \text{in} - t = 8.0\ \text{in} - 0.5\ \text{in} = 7.5\ \text{in}$$
$$e = \frac{3b^2}{h+6b} = \frac{(3)(3.75\ \text{in})^2}{7.5\ \text{in} + (6)(3.75\ \text{in})} = 1.41\ \text{in}$$

Since the applied load acts through the shear center, there is no torsion on the beam. The load is purely flexural.

The moment due to flexure is

$$M = \frac{PL}{4} = \frac{(200\ \text{lbf})(10\ \text{ft})}{4} = 500\ \text{ft-lbf (flexure)}$$

The answer is (D).

Why Other Options Are Wrong

(A) This incorrect solution does not check the shear center of the channel beam and assumes that the load creates a torsional moment. This solution also ignores the flexural effects of the load.

$$T = Pe = (200\ \text{lbf})(1.41\ \text{in}) = 282\ \text{in-lbf} \quad \big(280\ \text{in-lbf (torsion)}\big)$$

(B) This incorrect solution does not check the shear center of the channel beam and assumes that the load creates a torsional moment. This solution also ignores the flexural effects of the load and lists the units incorrectly.

(C) This incorrect solution does not check the shear center of the channel beam and assumes that the load creates a torsional moment in addition to the flexural moment.

SOLUTION 9

From the sum of the moments about E, the vertical reaction at support A is

$$R_{A,v} = \frac{(300\ \text{lbf})(2h) + (200\ \text{lbf})(3h)}{4h} = 300\ \text{lbf}$$

Draw the free-body diagram of joint A.

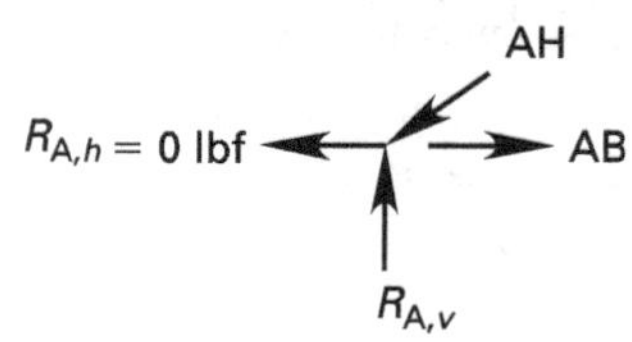

By summing the forces in the horizontal direction, determine that the horizontal reaction at A is zero.

The sum of the forces in the vertical direction is

$$\text{AH}_v - R_{A,v} = 0\ \text{lbf}$$
$$\text{AH}_v = 300\ \text{lbf (compression)}$$

From the geometry of the truss, the horizontal component of AH must be twice the vertical component.

$$\text{AH}_h = (2)(300\ \text{lbf}) = 600\ \text{lbf (compression)}$$

The resultant force in member AH is

$$\begin{aligned} \text{AH} &= \sqrt{\text{AH}_v^2 + \text{AH}_h^2} = \sqrt{(300\text{ lbf})^2 + (600\text{ lbf})^2} \\ &= 670.8\text{ lbf} \quad (670\text{ lbf (compression)}) \end{aligned}$$

The answer is (C).

Why Other Options Are Wrong

(A) This incorrect solution finds only the vertical component of the force in AH.

(B) This incorrect solution finds the horizontal component of the force in AH.

(D) This incorrect solution identifies the resultant force in AH as a tensile force.

SOLUTION 10

The free-body diagram for the beam described is shown.

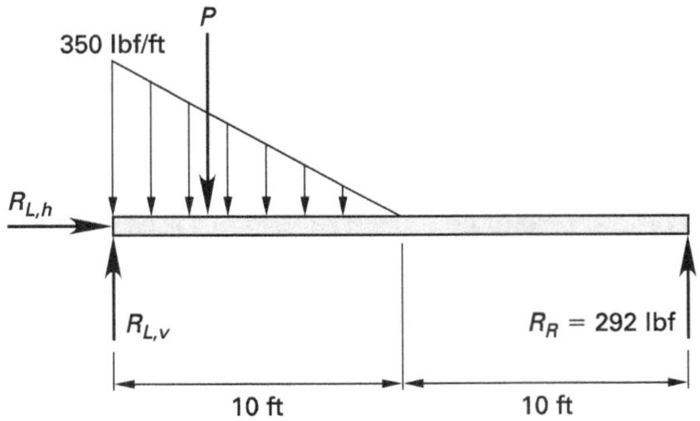

The equivalent point load for the triangular load is located at one-third the length of the load (one-third of 10 ft) and is

$$\begin{aligned} P &= \tfrac{1}{2}bh = \left(\frac{1}{2}\right)(10\text{ ft})\left(350\ \frac{\text{lbf}}{\text{ft}}\right) \\ &= 1750\text{ lbf} \end{aligned}$$

$$\begin{aligned} R_L &= P - R_R = 1750\text{ lbf} - 292\text{ lbf} \\ &= 1458\text{ lbf} \end{aligned}$$

The shear diagram for this beam is as follows.

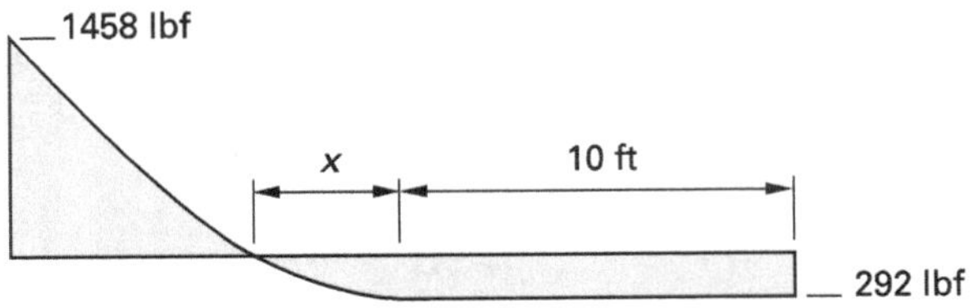

From the shear diagram, determine that the point of zero shear is between 0 and 10 ft from the left support. In this case, it is easier to write the equation for shear at a distance greater than 10 ft from the right support.

Shear is equal to the area under the load. The area under the load at a distance x from the base of the triangle is

$$A_x = \frac{1}{2}x\left(\frac{350\ \frac{\text{lbf}}{\text{ft}}}{10\text{ ft}}\right)x$$

The equation for shear at a distance x measured from greater than 10 ft from the right support is

$$\begin{aligned} V_x &= R_R - A_x = 292\text{ lbf} - \frac{1}{2}x\left(\frac{350\ \frac{\text{lbf}}{\text{ft}}}{10\text{ ft}}\right)x \\ &= 292\text{ lbf} - \left(\frac{1}{2}\right)\left(35\ \frac{\text{lbf}}{\text{ft}^2}\right)x^2 \end{aligned}$$

Shear equals 0 at

$$\begin{aligned} 0 &= 292\text{ lbf} - \left(17.5\ \frac{\text{lbf}}{\text{ft}^2}\right)x^2 \\ x &= 4.08\text{ ft} \quad [\text{measured left of center}] \end{aligned}$$

The distance to the point of zero shear, measured from the left end of the beam, is

$$\begin{aligned} D_{0\text{ shear},L} &= 10\text{ ft} - x = 10\text{ ft} - 4.08\text{ ft} \\ &= 5.92\text{ ft} \quad (5.9\text{ ft}) \end{aligned}$$

The answer is (C).

Why Other Options Are Wrong

(A) This incorrect solution writes the shear equation at a point measured from the left support but does not include the equivalent point load in the shear equation.

The equation for shear at a distance x from the left support is

$$\begin{aligned} V_x &= R_L - \left(\left(\frac{1}{2}\right)(10\text{ ft} - x)\left(\frac{350\ \frac{\text{lbf}}{\text{ft}}}{10\text{ ft}}\right)(10\text{ ft} - x)\right) \\ &= 1458\text{ lbf} - \left(\left(\frac{1}{2}\right)(10\text{ ft} - x)\left(\frac{350\ \frac{\text{lbf}}{\text{ft}}}{10\text{ ft}}\right)(10\text{ ft} - x)\right) \\ &= 1458\text{ lbf} - \left(17.5\ \frac{\text{lbf}}{\text{ft}^2}\right)(10\text{ ft} - x)^2 \end{aligned}$$

Shear equals 0 at

$$0 = 1458 \text{ lbf} - \left(17.5 \ \frac{\text{lbf}}{\text{ft}^2}\right)(10 \text{ ft} - x)^2$$
$$(10 \text{ ft} - x)^2 = 83.31 \text{ ft}^2$$
$$x = 0.87 \text{ ft}$$

(B) This incorrect solution calculates the distance not from the left support but rather from the center of the beam.

(D) This incorrect solution distributes the load over the entire beam instead of just one-half of it.

SOLUTION 11

Deflection is a function of the moment of inertia of the beam. Using the table of properties found in the AISC *Steel Construction Manual*, for a W16 × 31 beam,

$$I = 375 \text{ in}^4$$
$$E = 29 \times 10^6 \text{ lbf/in}^2$$
$$w_{\text{self-wt}} = 31 \text{ lbf/ft}$$

$$\begin{aligned} w &= w_D + w_L + w_{\text{self-wt}} \\ &= 500 \ \frac{\text{lbf}}{\text{ft}} + 1000 \ \frac{\text{lbf}}{\text{ft}} + 31 \ \frac{\text{lbf}}{\text{ft}} \\ &= 1531 \text{ lbf/ft} \end{aligned}$$

From the AISC *Steel Construction Manual* Table 3-23, find the three-span uniformly loaded beam in row 39.

From the problem statement, $L = 20$ ft.

Therefore, the maximum deflection in the first span is given as

$$\begin{aligned} \Delta_{\text{max}} &= \frac{0.0069 w L^4}{EI} \\ &= \frac{(0.0069)\left(1531 \ \frac{\text{lbf}}{\text{ft}}\right)(20 \text{ ft})^4\left(12 \ \frac{\text{in}}{\text{ft}}\right)^3}{\left(29 \times 10^6 \ \frac{\text{lbf}}{\text{in}^2}\right)(375 \text{ in}^4)} \\ &= 0.27 \text{ in} \end{aligned}$$

The answer is (B).

Why Other Options Are Wrong

(A) This incorrect solution approximates the deflection of the first span by using the loading diagram for a beam pinned at one end and fixed at the other from the AISC Table 3-23. Although a continuous-span end condition is sometimes approximated by a fixed end, the actual loading diagram should be used when available.

(C) When calculating the deflection, this incorrect solution uses the moment of inertia for a W16 × 26 beam instead of a W16 × 31 beam from the AISC Table 3-23.

(D) This incorrect solution does not consider that the spans are continuous. It finds the maximum deflection for a uniformly loaded beam with pinned supports.

SOLUTION 12

To find the centroid of the cross section, divide the section into two rectangles.

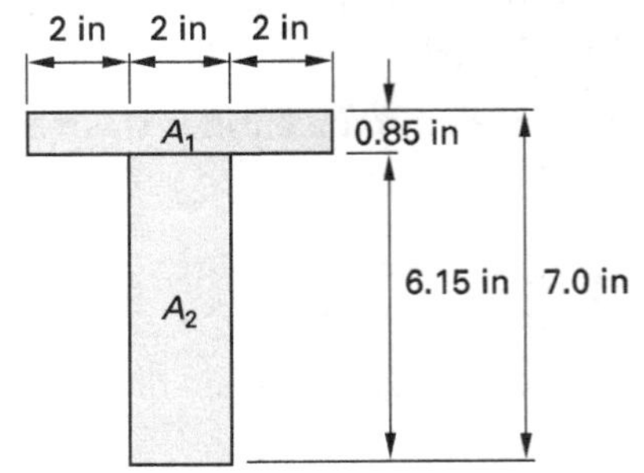

$$A_1 = (0.85 \text{ in})(6 \text{ in}) = 5.1 \text{ in}^2$$
$$A_2 = (6.15 \text{ in})(2 \text{ in}) = 12.3 \text{ in}^2$$

The distance from the top of the section to the centroid is

$$\begin{aligned} y_c &= \frac{\sum A_i y_{ci}}{\sum A_i} \\ &= \frac{(5.1 \text{ in}^2)\left(\frac{0.85 \text{ in}}{2}\right) + (12.3 \text{ in}^2)\left(\frac{6.15 \text{ in}}{2} + 0.85 \text{ in}\right)}{5.1 \text{ in}^2 + 12.3 \text{ in}^2} \\ &= 2.9 \text{ in} \quad [\text{from top of section}] \end{aligned}$$

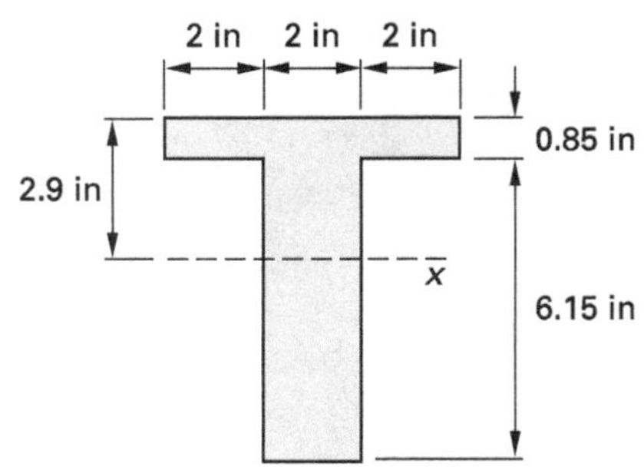

The shear stress is

$$\tau = \frac{VQ}{Ib}$$

Calculate I and Q. For a rectangular section,

$$I_1 = \frac{bh^3}{12} = \frac{(6 \text{ in})(0.85 \text{ in})^3}{12} = 0.31 \text{ in}^4$$

$$I_2 = \frac{bh^3}{12} = \frac{(2 \text{ in})(6.15 \text{ in})^3}{12} = 38.8 \text{ in}^4$$

The moment of inertia about the centroid is

$$\begin{aligned} I_x &= \sum (I_i + A_i d_i^2) \\ &= 0.31 \text{ in}^4 + (5.1 \text{ in}^2)\left(2.9 \text{ in} - \frac{0.85 \text{ in}}{2}\right)^2 + 38.8 \text{ in}^4 \\ &\quad + (12.3 \text{ in}^2)\left(7.0 \text{ in} - \frac{6.15 \text{ in}}{2} - 2.9 \text{ in}\right)^2 \\ &= 83.3 \text{ in}^4 \end{aligned}$$

The statical moment of the area, Q, is the product of the area above or below the point in question and the distance from the centroidal axis to the centroid of the area. Looking at the area below the centroid,

$$\begin{aligned} Q &= A\bar{y} = \tfrac{1}{2}b(h - y_c)^2 \\ &= \left(\frac{1}{2}\right)(2 \text{ in})(7.0 \text{ in} - 2.9 \text{ in})^2 \\ &= 16.8 \text{ in}^3 \end{aligned}$$

The shear stress at the centroid is

$$\begin{aligned} \tau &= \frac{VQ}{Ib} = \frac{(100 \text{ lbf})(16.8 \text{ in}^3)}{(83.3 \text{ in}^4)(2 \text{ in})} \\ &= 10.1 \text{ lbf/in}^2 \quad (10 \text{ lbf/in}^2) \end{aligned}$$

The answer is (C).

Why Other Options Are Wrong

(A) This incorrect solution miscalculates the location of the centroid of the section. The distance from the centroid of area A_2 to the top of the section fails to include the 0.85 in thickness of area A_1.

(B) This incorrect solution miscalculates A_2 and carries the mistake throughout the subsequent calculations. The length of A_2 is taken as the overall length of 7.0 in instead of the actual length of 6.15 in.

(D) This incorrect solution does not properly calculate the transformed moment of inertia. It directly adds the moments of inertia for each area instead of calculating the transformed moment of inertia.

SOLUTION 13

Find the rigidity of the walls. Rigidity is the relative stiffness of the walls and is proportional to the inverse of the shear and flexural deflection of the walls. However, for squat walls with h/L no more than 0.25, it is reasonably accurate to calculate the rigidity based on shear alone. For walls with h/L greater than 0.25 and less than 4, both flexure and shear must be considered. For very tall walls, the contribution from shear deformations is very small, and the rigidity can be based on flexure alone.

Since walls B, C, and E have the shortest length, the critical aspect ratio for these walls is

$$\frac{h}{L} = \frac{9 \text{ ft}}{36 \text{ ft}} = 0.25$$

Therefore, the rigidity can be based on shear alone. Assume the walls all have the same modulus of elasticity. Since the walls are all the same height, h_j, the rigidity is proportional to the area of the walls. If the walls had the same thickness, the shear rigidity would be proportional to wall length alone.

$$r_j = \frac{k_j}{\sum k_i}$$

$$k_j = \frac{A_j E_j}{h_j}$$

$$r_j = \frac{k_j}{k_j + k_k} = \frac{A_j E_j}{h_j\left(\dfrac{A_j E_j}{h_j} + \dfrac{A_k E_k}{h_k}\right)}$$

$$r_j \propto A = tL$$

$$r_A \propto (16 \text{ in})\left(\frac{1 \text{ ft}}{12 \text{ in}}\right)(60 \text{ ft}) = 80 \text{ ft}^2$$

$$r_B \propto (16 \text{ in})\left(\frac{1 \text{ ft}}{12 \text{ in}}\right)(36 \text{ ft}) = 48 \text{ ft}^2$$

$$r_C \propto (12 \text{ in})\left(\frac{1 \text{ ft}}{12 \text{ in}}\right)(40 \text{ ft}) = 40 \text{ ft}^2$$

$$r_D \propto (12 \text{ in})\left(\frac{1 \text{ ft}}{12 \text{ in}}\right)(36 \text{ ft}) = 36 \text{ ft}^2$$

$$r_E \propto (16 \text{ in})\left(\frac{1 \text{ ft}}{12 \text{ in}}\right)(36 \text{ ft}) = 48 \text{ ft}^2$$

Find the center of rigidity in each direction.

The distance to the center of rigidity from the south edge is

$$\bar{y} = \frac{\sum r_i y_i}{\sum r_i} = \frac{r_A y_A + r_D y_D}{r_A + r_D}$$
$$= \frac{(80 \text{ ft}^2)(160 \text{ ft}) + (36 \text{ ft}^2)(40 \text{ ft})}{80 \text{ ft}^2 + 36 \text{ ft}^2}$$
$$= 123 \text{ ft}$$

The distance to the center of rigidity from the west edge is

$$\bar{x} = \frac{\sum r_i x_i}{\sum r_i} = \frac{r_B x_B + r_C x_C + r_E x_E}{r_B + r_C + r_E}$$
$$= \frac{(48 \text{ ft}^2)(80 \text{ ft}) + (40 \text{ ft}^2)(50 \text{ ft}) + (48 \text{ ft}^2)(20 \text{ ft})}{48 \text{ ft}^2 + 40 \text{ ft}^2 + 48 \text{ ft}^2}$$
$$= 50.0 \text{ ft}$$

The answer is (D).

Why Other Options Are Wrong

(A) This incorrect solution does not include the effects of the different wall thicknesses when calculating the rigidities.

(B) This incorrect solution does not include the effects of the different wall thicknesses and makes a calculation error when determining the distance to the center of rigidity from the west end by calculating $\sum r_i$ equal to 102 ft instead of 112 ft.

(C) This incorrect solution correctly finds the center of rigidity but reverses the direction from which the distance is measured when choosing the answer.

SOLUTION 14

The shear force in the walls is a combination of the total shear force on the building and the shear induced by the torsional moment. If the resultant wind load, W, does not pass through the center of rigidity (c.r.) of the structure, the resultant creates a torsional moment that is resisted by shear in the walls.

$$V = \frac{W r_i}{\sum r_i} + \frac{M_t x r_i}{I_p}$$
$$W = wL = \left(200 \ \frac{\text{lbf}}{\text{ft}}\right)(160 \text{ ft})$$
$$= 32{,}000 \text{ lbf} \quad (32 \text{ kips})$$
$$I_p = I_{xx} + I_{yy}$$
$$I_{xx} = \sum A y^2$$
$$I_{yy} = \sum A x^2$$

Calculate I_{xx} using walls B and C. Calculate I_{yy} using walls A and D. x and y are the distances from the centroids of each wall area to the center of rigidity of the building.

$$I_{xx} = \sum A y^2$$
$$= (20 \text{ ft})(1 \text{ ft})(10 \text{ ft})^2 + (20 \text{ ft})(1 \text{ ft})(10 \text{ ft})^2$$
$$= 4000 \text{ ft}^4$$
$$I_{yy} = \sum A x^2$$
$$= (50 \text{ ft})(1 \text{ ft})(98.3 \text{ ft} - 65 \text{ ft})^2 + (40 \text{ ft})(1 \text{ ft})(140 \text{ ft} - 98.3 \text{ ft})^2$$
$$= 125{,}000 \text{ ft}^4$$
$$I_p = I_{xx} + I_{yy} = 4000 \text{ ft}^4 + 125{,}000 \text{ ft}^4$$
$$= 129{,}000 \text{ ft}^4$$

The torsional moment is the product of the resultant and the distance between the centroid of the load and the center of rigidity.

$$M_t = W\bar{x}$$
$$= (32 \text{ kips})(98.3 \text{ ft} - 80 \text{ ft})$$
$$= 586 \text{ ft-kips}$$

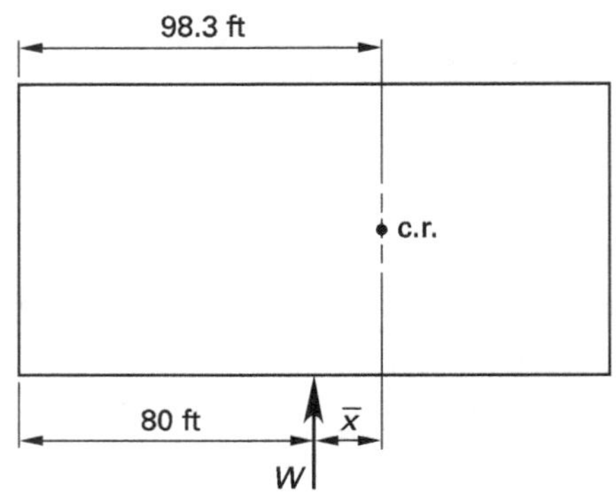

Rigidity is the relative stiffness of the walls. In this case, all walls are 12 in thick. Assume the walls all have the same modulus of elasticity and height, L_j.

$$r_j = \frac{k_j}{\sum k_i}$$
$$k_j = \frac{A_j E_j}{L_j}$$
$$r_j = \frac{k_j}{k_j + k_k} = \frac{A_j E_j}{L_j\left(\frac{A_j E_j}{L_j} + \frac{A_k E_k}{L_k}\right)}$$

Since E_j and E_k are equal and L_j and L_k are equal, the rigidity becomes

$$r_j = \frac{A_j}{A_j + A_k}$$

For wall A,

$$\begin{aligned} V_a &= \frac{Wr_a}{\sum r_i} + \frac{M_t x r_a}{I_p} \\ &= \frac{(32\ \text{kips})(50\ \text{ft}^2)}{50\ \text{ft}^2 + 40\ \text{ft}^2} \\ &\quad + \frac{(586\ \text{ft-kips})(98.3\ \text{ft} - 65\ \text{ft})(50\ \text{ft}^2)}{129{,}000\ \text{ft}^4} \\ &= 25.3\ \text{kips} \quad (25\ \text{kips}) \end{aligned}$$

The answer is (D).

Why Other Options Are Wrong

(A) This incorrect solution finds the shear load on wall D instead of wall A.

(B) This incorrect solution includes the east and west walls in calculating rigidity and does not include the effects of the eccentricity of the resultant (torsional moment).

For wall A,

$$\begin{aligned} V_a &= \frac{Wr_a}{\sum r_i} = \frac{(32\ \text{kips})(50\ \text{ft}^2)}{50\ \text{ft}^2 + 20\ \text{ft}^2 + 20\ \text{ft}^2 + 40\ \text{ft}^2} \\ &= 12.3\ \text{kips} \quad (12\ \text{kips}) \end{aligned}$$

(C) This incorrect solution does not include the effects of the torsional moment.

SOLUTION 15

The degree of indeterminacy of a pin-connected truss is given by the equation

$$\begin{aligned} \text{degree of indeterminacy} = 3 &+ \text{number of members} \\ &- 2(\text{number of joints}) \end{aligned}$$

In this case, there is one degree of indeterminacy or redundancy.

$$\text{degree of indeterminacy} = 3 + 6 - (2)(4) = 1$$

The forces in an indeterminate truss cannot be solved directly. Since there is only one redundant member, use the dummy unit-load method to determine the force in member BC.

step 1: Draw the truss twice. Omit the redundant member on both trusses.

step 2: Load the first truss (which is now determinate) with the actual loads.

step 3: Calculate the forces, S, in all of the members. Use a positive sign for tensile forces.

step 4: Load the second truss with two unit forces acting collinearly toward each other along the line of the redundant member.

step 5: Calculate the force, u, in each of the members.

step 6: Calculate the force in the redundant member using the equation

$$S_{\text{redundant}} = \frac{-\sum \frac{SuL}{AE}}{\sum \frac{u^2 L}{AE}}$$

step 7: The true force in member j of the truss is

$$F_{j,\text{true}} = S_j + S_{\text{redundant}} u_j$$

Removing the redundant member, draw the free-body diagram of the loaded truss. The reaction loads are calculated by static analysis of the truss.

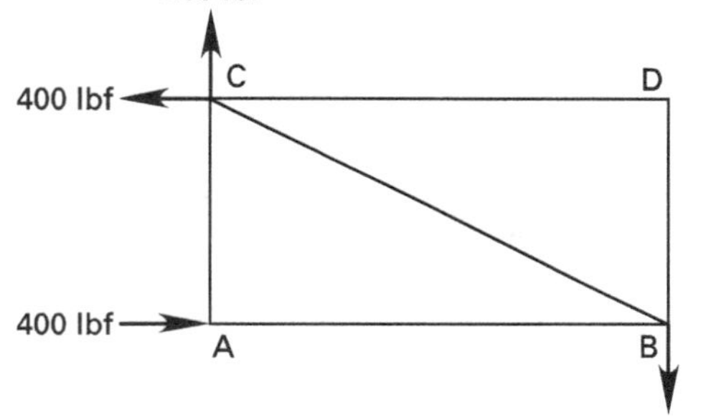

Draw the free-body diagram for each joint to determine the force in the members.

Joint A:

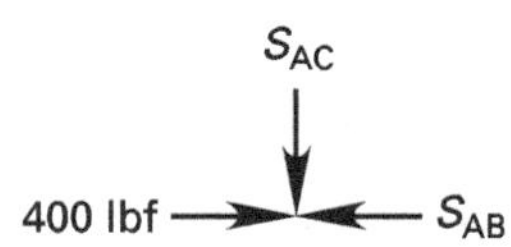

Summing the forces,

$$\begin{aligned} S_{\text{AC}} &= 0 \\ S_{\text{AB}} &= -400\ \text{lbf} \end{aligned}$$

Joint B:

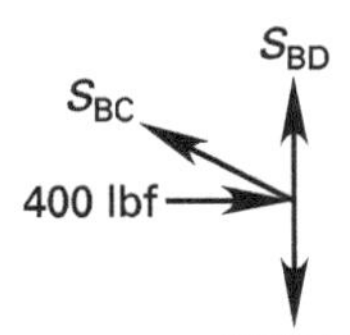

Summing the forces, the horizontal component of the force in member BC must equal 400 lbf (tension). By geometry of the figure, determine that the vertical component of member BC equals 200 lbf. Therefore, the force in member BD is zero. The resultant force in member BC is

$$S_{BC} = \sqrt{(400 \text{ lbf})^2 + (200 \text{ lbf})^2} = 447 \text{ lbf}$$

Continue in the same fashion, solving for the forces in the truss.

member	L (ft)	AE (kips)	S (lbf)	u	$\frac{SuL}{AE}$ (lbf-ft/kip)	$\frac{u^2L}{AE}$ (ft/kip)
AB	10	3	−400	−0.894	1192	2.66
AC	5	3	0	−0.447	0	0.33
CD	10	3	0	−0.894	0	2.66
BC	11.18	5	447	1.0	999	2.24
BD	5	3	0	−0.447	0	0.33
AD	11.18	5	0	1.0	0	2.24
					2191	10.46

Draw the free-body diagram of the unit-load truss, and solve for the member forces.

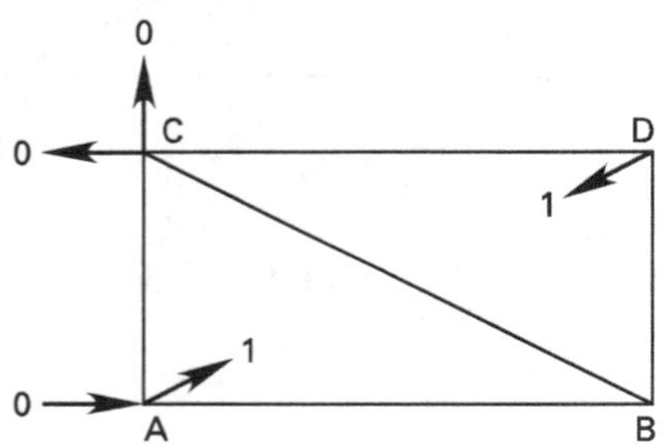

The force in redundant member AD is

$$S_{AD} = \frac{-\sum \frac{SuL}{AE}}{\sum \frac{u^2L}{AE}} = \frac{-2191 \frac{\text{lbf-ft}}{\text{kip}}}{10.46 \frac{\text{ft}}{\text{kip}}}$$
$$= -209 \text{ lbf}$$

The true force in member BC of the truss is

$$\begin{aligned} F_{BC,\text{true}} &= S_{BC} + S_{\text{redundant}}u_{BC} \\ &= 447 \text{ lbf} + (-209 \text{ lbf})(1.0) \\ &= 238 \text{ lbf (tension)} \quad (240 \text{ lbf (tension)}) \end{aligned}$$

The answer is (C).

Why Other Options Are Wrong

(A) This incorrect solution calculates the force in redundant member AD instead of the force in member BC.

(B) This incorrect solution uses the same product of area and modulus of elasticity, AE, for all members of the truss.

(D) This incorrect solution calculates the force in member BC in the determinate truss due to the applied load, S_{BC}, instead of the true force in member BC, $F_{BC,\text{true}}$.

SOLUTION 16

Find the reactions.

Member AB:

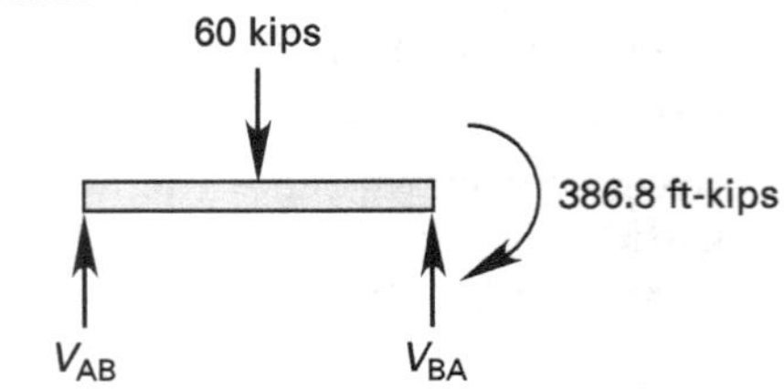

Taking clockwise moments and upward forces as positive,

$$\begin{aligned} \sum M_A &= 0 \text{ ft-kips} + (60 \text{ kips})(10 \text{ ft}) \\ &\quad +386.8 \text{ ft-kips} - V_{BA}(20 \text{ ft}) = 0 \\ V_{BA} &= 49.3 \text{ kips} \end{aligned}$$

Member BC:

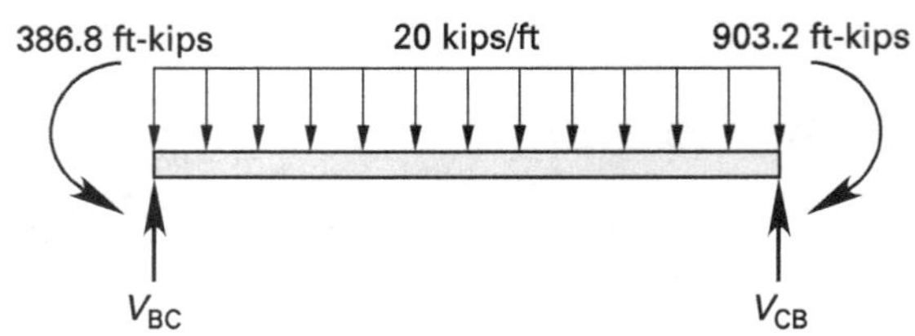

Taking clockwise moments and upward forces as positive,

$$\begin{aligned} \sum M_C &= 903.2 \text{ ft-kips} - \left(20 \frac{\text{kips}}{\text{ft}}\right)(20 \text{ ft})(10 \text{ ft}) \\ &\quad -386.8 \text{ ft-kips} + V_{BC}(20 \text{ ft}) \\ &= 0 \\ V_{BC} &= 174.2 \text{ kips} \end{aligned}$$

The reaction at B is the sum of the shears at B.

$$\begin{aligned} R_B &= V_{BA} + V_{BC} \\ &= 49.3 \text{ kips} + 174.2 \text{ kips} \\ &= 223.5 \text{ kips} \quad (220 \text{ kips}) \end{aligned}$$

The answer is (D).

Why Other Options Are Wrong

(A) This incorrect solution only considers the shear to the left of B in determining the reaction at B instead of including the shear from BC.

(B) This incorrect solution subtracts the shear forces on each side of B rather than adding them to determine the reaction at the support.

(C) This incorrect solution only considers the shear to the right of B in determining the reaction at B instead of including the shear from BA as well.

SOLUTION 17

Section 8.4.15 of ACI 318 defines a column strip as a width on each side of a column centerline equal to $0.25l_2$ or $0.25l_1$, whichever is less. A middle strip is the strip bound by two column strips.

l_1 is defined as the length of span in the direction in which moments are being determined, measured center-to-center of the supports.

l_2 is the length of span transverse to l_1, measured center-to-center of the supports.

l_n is the length of clear span in the direction that moments are being determined, measured face-to-face of the supports.

The midspan moment of the middle strip will be a positive moment. To find the moment in the middle strip, first determine the column strip moment.

The critical factored load on the slab, w_u, is 168 lbf/ft^2.

The total factored static moment on the slab given in ACI 318 Sec. 8.10.6.2 is

$$l_1 = 40 \text{ ft}$$
$$l_2 = 30 \text{ ft}$$

The clear span of the slab panel is

$$l_n = 40 \text{ ft} - (2)\left(\frac{1 \text{ ft}}{2}\right) = 39 \text{ ft}$$

$$\begin{aligned} M_o &= \frac{w_u l_2 l_n^2}{8} \\ &= \frac{\left(168 \; \frac{\text{lbf}}{\text{ft}^2}\right)\left(1000 \; \frac{\text{lbf}}{\text{kip}}\right)(30 \text{ ft})(39 \text{ ft})^2}{8} \\ &= 958 \text{ ft-kips} \end{aligned}$$

Per ACI Sec. 8.10.4, the total static moment, M_o, is distributed as

$$-M_u = -0.65M_o$$
$$M_u = 0.35M_o$$

Since the midspan moment is positive, use the factored positive moment to find the positive moment in the column and middle strips.

$$\begin{aligned} M_u &= 0.35M_o \\ &= (0.35)(958 \text{ ft-kips}) \\ &= 335 \text{ ft-kips} \end{aligned}$$

ACI 318 Sec. 8.10.5.5 gives the percentage of positive factored moment distributed to the column strips based on the ratio of the stiffness of the beam to the stiffness of the slab, $\alpha_{f1}l_2/l_1$. Since this is a flat plate, there are no beams and α_{f1} is zero.

$$\frac{\alpha_{f1}l_2}{l_1} = \frac{(0)(30 \text{ ft})}{40 \text{ ft}} = 0$$

Use ACI 318 Table 8.10.5.5 to find the percentage of positive factored moment distributed to the column strips to be 60%.

Since the columns are equally spaced on each side of the slab strip, the percentage of positive moment in the middle strip is

$$\begin{aligned} M_{\text{middle},\%} &= 100\% - 60\% \\ &= 40\% \end{aligned}$$

The positive moment in the middle strip is

$$\begin{aligned} M_{\text{middle}} &= (40\%)M_u \\ &= (0.40)(335 \text{ ft-kips}) \\ &= 134 \text{ ft-kips} \quad (130 \text{ ft-kips}) \end{aligned}$$

The answer is (C).

Why Other Options Are Wrong

(A) This incorrect solution uses α_{f1} equal to 1.0 instead of 0 when calculating the percentage of positive moment distributed to the column strips.

(B) This incorrect solution reverses the values of l_1 and l_2 in calculating the total factored static moment on the slab, M_o. l_1 is in the direction in which the moments are being determined.

(D) This incorrect solution uses the center-to-center span length instead of the clear span, l_n, in determining the total factored static moment.

SOLUTION 18

Use the AISC *Steel Construction Manual* (AISC *Manual*) Table 3-23 to determine the deflection of the beam. The deflection for the loads given is the sum of the loads taken individually at the specified point.

The deflection of a uniform load partially distributed at one end at a point 5 ft from the left support of the beam is given in AISC *Manual* Table 3-23, diagram 5, as

$$\begin{aligned}\Delta_{w,x=5\text{ ft}} &= \frac{wa^2}{24EIl}(l-x)(4xl-2x^2-a^2)\\ &= \left(\frac{\left(20\ \frac{\text{lbf}}{\text{ft}}\right)(5\text{ ft})^2}{(24)\left(1.2\times10^6\ \frac{\text{lbf}}{\text{in}^2}\right)(240\text{ in}^4)(11\text{ ft})}\right)\\ &\quad\times(11\text{ ft}-5\text{ ft})\\ &\quad\times\left((4)(5\text{ ft})(11\text{ ft})-(2)(5\text{ ft})^2-(5\text{ ft})^2\right)\\ &\quad\times\left(12\ \frac{\text{in}}{\text{ft}}\right)^3\\ &= 0.0099\text{ in}\end{aligned}$$

From AISC *Manual* Table 3-23, diagram 8, the deflection of a point load at a point 5 ft from the left support of the beam is

$$\begin{aligned}\Delta_{P,x=5\text{ ft}} &= \frac{Pbx}{6EIl}(l^2-b^2-x^2)\\ &= \left(\frac{(100\text{ lbf})(3\text{ ft})(5\text{ ft})}{(6)\left(1.2\times10^6\ \frac{\text{lbf}}{\text{in}^2}\right)(240\text{ in}^4)(11\text{ ft})}\right)\\ &\quad\times\left((11\text{ ft})^2-(3\text{ ft})^2-(5\text{ ft})^2\right)\\ &\quad\times\left(12\ \frac{\text{in}}{\text{ft}}\right)^3\\ &= 0.012\text{ in}\\ \Delta_{\text{total}} &= \Delta_w+\Delta_P = 0.0099\text{ in}+0.012\text{ in}\\ &= 0.0219\text{ in}\quad(0.022\text{ in})\end{aligned}$$

The answer is (D).

Why Other Options Are Wrong

(A) This incorrect solution only calculates the deflection due to the uniform load and neglects the deflection due to the point load.

(B) This incorrect solution calculates the deflection from the point load at a distance 8 ft from the left support (at the location of the point load) and adds it to the deflection from the uniform load at a distance 5 ft from the left support.

(C) This incorrect solution makes a mistake in the unit conversion when calculating the deflection of the uniform load.

SOLUTION 19

The torsional moment on the beam restrained at one end is

$$\begin{aligned}T = wLe &= \left(20\ \frac{\text{kips}}{\text{ft}}\right)(20\text{ ft})(9\text{ in})\left(\frac{1\text{ ft}}{12\text{ in}}\right)\\ &= 300\text{ ft-kips}\end{aligned}$$

The answer is (B).

Why Other Options Are Wrong

(A) This incorrect solution calculates the torsion for a beam restrained at both ends.

(C) This incorrect solution calculates the flexural moment instead of the torsional moment for a beam with both ends fully restrained (fixed).

(D) This incorrect solution calculates the flexural moment instead of the torsional moment for a simply supported beam without end restraint.

SOLUTION 20

Spandrel beams AB, BC, and CD carry a uniform dead and live load based on a tributary load width of $(10\text{ ft})/2 = 5$ ft. Spandrel beams AE and DH carry concentrated loads due to the beams framing into them with an equivalent tributary load width of $(5\text{ ft}+10\text{ ft}+5\text{ ft})/2 = 10$ ft. The live load on AE is twice the live load on AB. All spandrel beams also carry the dead load due to the brick veneer.

The answer is (C).

Why Other Options Are Wrong

(A) This incorrect solution does not take into account that though the tributary area for interior beam FG is twice that of beam BC, beam BC also carries the dead load of the brick veneer.

(B) This incorrect solution does not take into account that the floor loads on beam AE are 2 times the floor loads on BC, and both carry an equal dead load due to the brick veneer.

(D) This incorrect solution does not take into account that the floor loads on beam AE are 2 times the floor loads on AB, and both carry an equal dead load due to the brick veneer.

SOLUTION 21

The *International Building Code* (IBC) Sec. 1608 refers to *Minimum Design Loads for Buildings and Other Structures* (ASCE/SEI7) Chap. 7 for determining the snow exposure factor, C_e. ASCE/SEI7 gives C_e as a function of the roof exposure and terrain categories.

Given that the buildings in the office park have the same footprint and height, treat them as having the same exposure.

From ASCE/SEI7 Table 7-2, a building whose roof is not sheltered by terrain, higher structures, or trees is categorized as fully exposed. The office park is flat, all buildings are of the same height, and only low-level ground cover is used for landscaping. Therefore, the building may be categorized as fully exposed.

From ASCE/SEI7 Sec. 26.7, the terrain category is a function of the surface roughness and exposure. The office park is in a suburban area, so it has a surface roughness category of B (ASCE/SEI7 Sec. 26.7.2). The buildings have heights greater than 30 ft and are located in a 100 acre office park (i.e., the surface roughness B will extend more than 2600 ft), so the terrain exposure category is B (ASCE/SEI7 Sec. 26.7.3).

From ASCE/SEI7 Table 7-2, the exposure factor for a fully exposed building with a terrain category of B is 0.9.

The answer is (B).

Why Other Options Are Wrong

(A) This incorrect solution uses terrain exposure category D instead of B.

(C) This incorrect solution finds the correct terrain exposure category, but finds the snow exposure factor for a partially exposed roof instead of a fully exposed roof.

(D) This incorrect solution finds the correct terrain exposure category, but finds the snow exposure factor for a sheltered roof instead of a fully exposed roof.

SOLUTION 22

Calculate the number of bolts per joist.

$$n = \frac{\text{joist spacing}}{\text{bolt spacing}} = \frac{6.0 \text{ ft}}{(16 \text{ in})\left(\frac{1 \text{ ft}}{12 \text{ in}}\right)} = 4.5 \text{ bolts/joist}$$

Calculate the loads per bolt.

The applied shear load per bolt is

$$b_v = \frac{R}{n} = \frac{3700 \ \frac{\text{lbf}}{\text{joist}}}{4.5 \ \frac{\text{bolts}}{\text{joist}}} = 822 \text{ lbf/bolt} \quad (820 \text{ lbf/bolt})$$

The prying tension per joist is

$$T = \frac{Rx}{y} = \frac{\left(3700 \ \frac{\text{lbf}}{\text{joist}}\right)(2.5 \text{ in})}{3 \text{ in}} = 3083 \text{ lbf/joist}$$

The prying tension per bolt is

$$b_a = \frac{T}{n} = \frac{3083 \ \frac{\text{lbf}}{\text{joist}}}{4.5 \ \frac{\text{bolts}}{\text{joist}}} = 685 \text{ lbf/bolt} \quad (690 \text{ lbf/bolt})$$

The load on each anchor bolt is 820 lbf in shear and 690 lbf in tension.

The answer is (C).

Why Other Options Are Wrong

(A) This incorrect solution makes an error in the unit conversion when calculating the number of bolts per joist.

(B) This incorrect solution neglects the prying tension on the anchors.

(D) This incorrect solution doesn't distribute the load based on the spacing of the joists and anchor bolts.

SOLUTION 23

The tributary width for an interior column is found by summing the halved distances to the adjacent column(s) from the column centerline.

$$w' = \sum \tfrac{1}{2}d$$

$$w_1' = \left(\frac{1}{2}\right)(20 \text{ ft}) + \left(\frac{1}{2}\right)(20 \text{ ft}) = 20 \text{ ft}$$

$$w_2' = \left(\frac{1}{2}\right)(20 \text{ ft}) + \left(\frac{1}{2}\right)(20 \text{ ft}) = 20 \text{ ft}$$

The tributary area for an interior column is

$$A = w_1'w_2' = (20\ \text{ft})(20\ \text{ft}) = 400\ \text{ft}^2$$

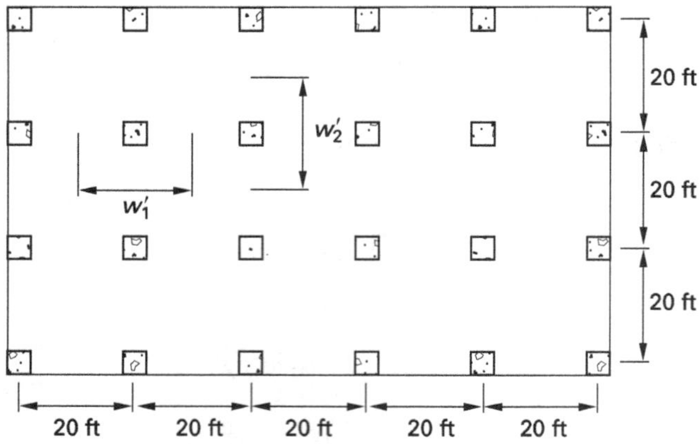

From IBC Sec. 1605.3.1, the applicable basic load combinations for allowable stress design are

$$\begin{gathered} D + L_{\text{floor}} \\ D + L_{\text{roof}} \\ D + 0.75(L_{\text{floor}} + L_{\text{roof}}) \\ 0.6D \end{gathered}$$

By inspection, $0.6D$ does not control.

The total uniform load per unit area is the larger of

$$\begin{aligned} w_{\text{uniform}} &= (D_{\text{roof}} + D_{\text{floor}}) + L_{\text{floor}} \\ &= \left(15\ \frac{\text{lbf}}{\text{ft}^2} + 15\ \frac{\text{lbf}}{\text{ft}^2}\right) + 60\ \frac{\text{lbf}}{\text{ft}^2} \\ &= 90\ \text{lbf/ft}^2 \\ w_{\text{uniform}} &= (D_{\text{roof}} + D_{\text{floor}}) + L_{\text{roof}} \\ &= \left(15\ \frac{\text{lbf}}{\text{ft}^2} + 15\ \frac{\text{lbf}}{\text{ft}^2}\right) + 20\ \frac{\text{lbf}}{\text{ft}^2} \\ &= 50\ \text{lbf/ft}^2 \\ w_{\text{uniform}} &= (D_{\text{roof}} + D_{\text{floor}}) + 0.75(L_{\text{floor}} + L_{\text{roof}}) \\ &= \left(15\ \frac{\text{lbf}}{\text{ft}^2} + 15\ \frac{\text{lbf}}{\text{ft}^2}\right) + >(0.75)\left(60\ \frac{\text{lbf}}{\text{ft}^2} + 20\ \frac{\text{lbf}}{\text{ft}^2}\right) \\ &= 90\ \text{lbf/ft}^2 \end{aligned}$$

The total column load is

$$\begin{aligned} P = w_{\text{uniform}}A &= \left(90\ \frac{\text{lbf}}{\text{ft}^2}\right)(400\ \text{ft}^2) \\ &= 36{,}000\ \text{lbf} \quad (36\ \text{kips}) \end{aligned}$$

The answer is (C).

Why Other Options Are Wrong

(A) This incorrect solution uses tributary width instead of tributary area when calculating the total load on the column.

(B) This incorrect solution does not include the second-floor loads when calculating the total load on the column.

(D) This incorrect solution does not consider the applicable load combinations and instead uses the total roof and floor live and dead loads.

SOLUTION 24

The load on the beam, w, is the weight of the wall per foot. Use NMCA TEK 14-13B to determine that a 6 in ungrouted hollow wall made from 85 lbf/ft^3 concrete has a weight of 20 lbf/ft^2. (The weight can also be determined by interpolating ASCE/SEI7 Table C3-1.) The load on the beam is

$$\begin{aligned} w = w_{\text{wall}}h &= \left(20\ \frac{\text{lbf}}{\text{ft}^2}\right)(8\ \text{ft}) \\ &= 160\ \text{lbf/ft} \end{aligned}$$

The answer is (B).

Why Other Options Are Wrong

(A) This incorrect solution fails to multiply by the height of the wall.

(C) This incorrect solution uses the column in NCMA TEK 14-13B for an 8 in wall.

(D) This incorrect solution calculates the weight of the wall as if solid or fully grouted.

SOLUTION 25

To determine what loads need to be considered on a lintel, first determine if arching action will occur. A masonry wall will exhibit arching action over an opening if sufficient masonry extends on both sides of the opening to resist the thrusting action of the arch and if there is sufficient wall height above the opening to permit the formation of a symmetrical 45° triangle plus 8 in as shown. An extent of masonry on each side greater than the distance of the opening itself is usually adequate to resist the thrust. In this case, arching action can be assumed to occur. When arching action occurs, only the loads within a 45° triangle over the opening will be carried by the lintel.

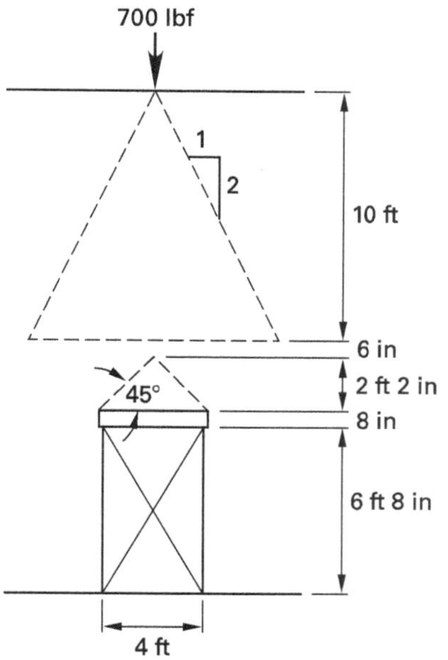

Section 5.1.3.1 of *Building Code Requirements for Masonry Structures* (TMS 402) requires that for walls laid in running bond, concentrated loads must be distributed over a length equal to the length of bearing plus the length determined by dispersing the concentrated load along a 2 vertical:1 horizontal line as shown. Because the dispersion cannot exceed half the wall height (10 ft), the concentrated roof load does not fall within the arching triangle of the lintel and does not need to be considered in this design.

The answer is (C).

Why Other Options Are Wrong

(A) This incorrect solution ignores the effect of the arching action of the wall.

(B) This incorrect solution ignores the arching action of the wall and incorrectly adds the effects of the full concentrated load.

(D) This incorrect solution considers the arching action of the wall but incorrectly adds the effects of the full concentrated load.

SOLUTION 26

The weight of the brick veneer is given as 40 lbf/ft^2. Brick veneer is typically supported at each floor. Therefore, the brick load on the spandrel beam BC is for a one-story height, or 13 ft. The spacing between beams is 10 ft. The tributary width, w', for a spandrel beam is

$$w' = \left(\frac{1}{2}\right)(10\ \text{ft}) = 5\ \text{ft}$$

Determine the loads on the spandrel beam.

For the dead load,

$$\begin{aligned}
w_{\text{floor,DL}} &= (5\ \text{ft})\left(60\ \frac{\text{lbf}}{\text{ft}^2}\right) \\
&= 300\ \text{lbf/ft} \\
w_{\text{self-wt}} &= 50\ \text{lbf/ft} \\
w_{\text{brick}} &= \left(40\ \frac{\text{lbf}}{\text{ft}^2}\right)(13\ \text{ft}) \\
&= 520\ \text{lbf/ft} \\
w_{\text{DL,total}} &= 300\ \frac{\text{lbf}}{\text{ft}} + 50\ \frac{\text{lbf}}{\text{ft}} + 520\ \frac{\text{lbf}}{\text{ft}} \\
&= 870\ \text{lbf/ft}
\end{aligned}$$

For the live load,

$$\begin{aligned}
w_{\text{floor,LL}} &= (5\ \text{ft})\left(40\ \frac{\text{lbf}}{\text{ft}^2}\right) \\
&= 200\ \text{lbf/ft}
\end{aligned}$$

Determine the controlling load combination.

Case 1: $1.2D + 1.6L = 1.2w_{\text{DL}} + 1.6w_{\text{LL}}$

$$\begin{aligned}
w_u &= 1.2w_{\text{DL,total}} + 1.6w_{\text{floor,LL}} \\
&= (1.2)\left(870\ \frac{\text{lbf}}{\text{ft}}\right) + (1.6)\left(200\ \frac{\text{lbf}}{\text{ft}}\right) \\
&= 1364\ \text{lbf/ft}
\end{aligned}$$

Case 2: $1.4D = 1.4w_{\text{DL}}$

$$\begin{aligned}
w_u &= 1.4w_{\text{DL,total}} = (1.4)\left(870\ \frac{\text{lbf}}{\text{ft}}\right) \\
&= 1218\ \text{lbf/ft}
\end{aligned}$$

Case 1 controls.

The maximum moment on the beam is

$$\begin{aligned}
M_{u,\max} &= \frac{w_u L^2}{8} = \frac{\left(1364\ \frac{\text{lbf}}{\text{ft}}\right)(30\ \text{ft})^2}{(8)\left(1000\ \frac{\text{lbf}}{\text{kip}}\right)} \\
&= 153\ \text{ft-kips}
\end{aligned}$$

The answer is (D).

Why Other Options Are Wrong

(A) This incorrect solution does not include the weight of the veneer when determining the design moment.

(B) This incorrect solution does not use any load factors when determining the design moment.

(C) This incorrect solution uses an incorrect load combination ($1.4D$) when determining the design moment.

2 Codes and Construction

CODES, STANDARDS, AND GUIDANCE DOCUMENTS

PROBLEM 1

Which of the following statements are true regarding reinforced concrete slabs designed using ACI 318?

I. The direct design method applies only to two-way slab systems.

II. The direct design method can be used for slabs with five continuous spans in one direction and two continuous spans in the other direction.

III. The number and length of spans is not restricted using the equivalent frame method of slab design.

IV. A slab with a dead load of 50 lbf/ft^2 and a live load of 150 lbf/ft^2 cannot be designed using the direct design method but can be designed using the equivalent frame method.

(A) I and II

(B) I and III

(C) I, III, and IV

(D) I, II, III, and IV

Hint: Refer to ACI 318 Chap. 8.

PROBLEM 2

A retail shopping center located in Utah is built with concrete masonry bearing walls and ordinary reinforced concrete masonry shear walls and is founded on soil. The walls of the shopping center are 15 ft tall. The maximum considered earthquake ground motion for 0.2 sec spectral response acceleration is 30% g. Using the simplified analysis procedure for seismic design found in ASCE/SEI7, the seismic base shear is

(A) $0.07W$

(B) $0.14W$

(C) $0.15W$

(D) $14W$

Hint: Refer to ASCE/SEI7 Sec. 12.14 for simplified seismic design.

PROBLEM 3

An interior bay of a two-way flat-slab system is supported by 12 in concrete columns as shown.

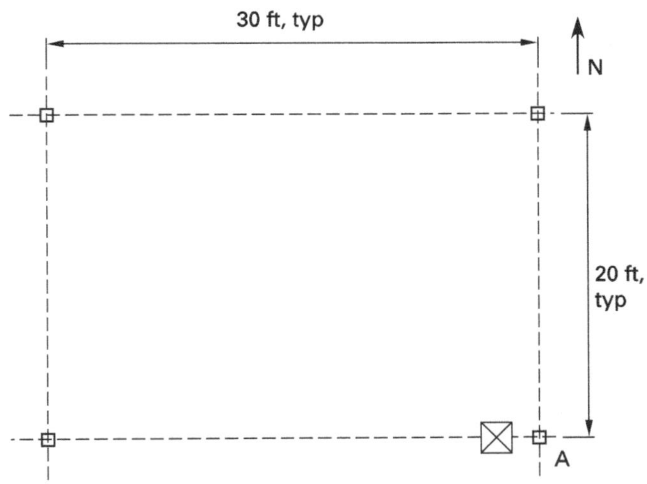

If no special analysis is used, the maximum size square opening that can be located adjacent to column A, centered on the east/west column centerline, is most nearly

(A) 0.0 ft^2

(B) 0.39 ft^2

(C) 1.6 ft^2

(D) 2.3 ft^2

Hint: Refer to ACI 318 Chap. 8 on two-way slabs.

PROBLEM 4

A single-story steel-framed building has columns spaced 15 ft on center in the north/south direction and 20 ft on center in the east/west direction. The columns support a uniform dead load of 40 lbf/ft². The building has an ordinary roof with a 1:2 pitch. Using the *International Building Code* (IBC), the minimum live load on an interior column is most nearly

(A) 4.1 kips

(B) 4.9 kips

(C) 6.0 kips

(D) 17 kips

Hint: IBC Sec. 1607 covers live loads.

PROBLEM 5

A two-story wood-framed apartment building is classified as seismic design category C according to the *International Building Code* (IBC). Which of the following statements about this building is FALSE?

(A) The total design lateral seismic force increases as the building weight increases.

(B) The short-period response accelerations for this site must be between 0.33 g and 0.50 g.

(C) There is no limit on story drift.

(D) When soil properties are not known in sufficient detail to determine the site class, site class D should be used unless determined otherwise by the building official or unless geotechnical data indicates that site class E or F soil is likely to be present.

Hint: Refer to IBC Sec. 1613.

PROBLEM 6

A church is built with a wood roof that is exposed on the interior. The roof beam is 60 ft in length. According to the *International Building Code* (IBC), what is the total deflection limit for the roof beam?

(A) 0.5 in

(B) 3 in

(C) 4 in

(D) 6 in

Hint: Refer to IBC Sec. 1604.3.

PROBLEM 7

Which of the following statements is/are FALSE regarding the *National Design Specification for Wood Construction* (NDS)?

I. The tabulated allowable bending design values include the effects of the beam stability factor.

II. The tabulated allowable design values for structural glued laminated timber (glulam) include adjustments for size.

III. The tabulated allowable design values for round timber piles include adjustments to compensate for the strength reduction associated with treated piles.

IV. The group action factor applies to all types of wood connections (i.e., nails, bolts, and spikes).

(A) I only

(B) II only

(C) I and IV

(D) I, III, and IV

Hint: Use the applicability tables in the NDS to determine which factors apply.

PROBLEM 8

A two-story, 40 ft by 60 ft building is located in a 100 acre office park in suburban Michigan. The building has an asphalt-shingle hip roof with a 6:12 pitch. The roof is supported by rafters spanning the short direction of the building from the eave to the ridge. The attic is vented, the floor is insulated with R-30 insulation, and the following load and factors apply.

snow importance factor, I_s	1.0
snow exposure factor, C_e	0.9
ground snow load, p_g	40 lbf/ft²

According to the *International Building Code* (IBC), the maximum leeward snow load is most nearly

(A) 0 lbf/ft²

(B) 20 lbf/ft²

(C) 30 lbf/ft²

(D) 40 lbf/ft²

Hint: Refer to ASCE/SEI7 Chap. 7.

PROBLEM 9

Which of the following seismic design statements are true?

I. The seismic base shear of a building increases as the building dead load increases.

II. In areas of high seismic activity, it is best to design buildings with an irregular floor plan to better break up the seismic load.

III. According to ASCE/SEI7, flat roof snow loads under 30 lbf/ft^2 need not be included in the effective seismic weight of the structure, W.

IV. Buildings with a soft first story and heavy roofs performed well during the Northridge earthquake in 1994.

(A) I and II only

(B) I and III only

(C) II and IV

(D) I, II, and III

Hint: Chapter 12 of ASCE/SEI7 provides information on seismic design.

PROBLEM 10

A 20 in wide × 30 in deep reinforced normalweight concrete beam has five continuous 20 ft spans. The beam is carrying uniform gravity loads such that the point of inflection for positive moment is 3 ft from the right face of the support. The minimum clear cover on the galvanized reinforcing bars is 1.5 in, and the minimum clear spacing of the bars is 2.5 in. Appropriate shear reinforcement is provided.

concrete compressive strength	6000 lbf/in^2
yield stress of reinforcement	60,000 lbf/in^2

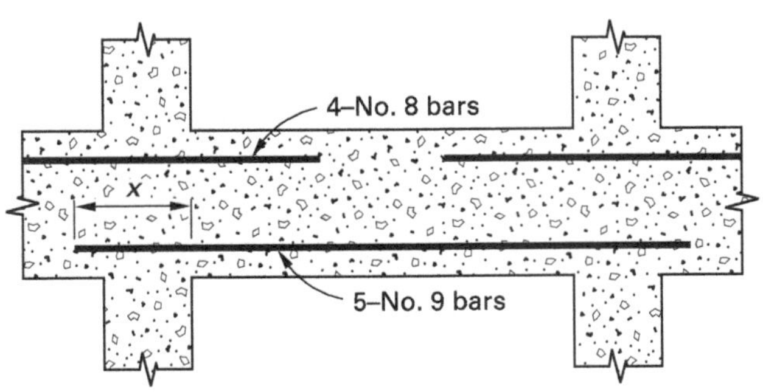

The required minimum length of the bottom bars beyond the face of the support, x, is most nearly

(A) 0 in

(B) 6 in

(C) 30 in

(D) 40 in

Hint: Refer to ACI 318 Chap. 25.

PROBLEM 11

A 20 ft diameter, smooth, circular water tank with a projected area of 315 ft^2 provides drinking water to a seaside resort town. The velocity pressure on the tank is 25 lbf/ft^2, and the gust factor is 0.85.

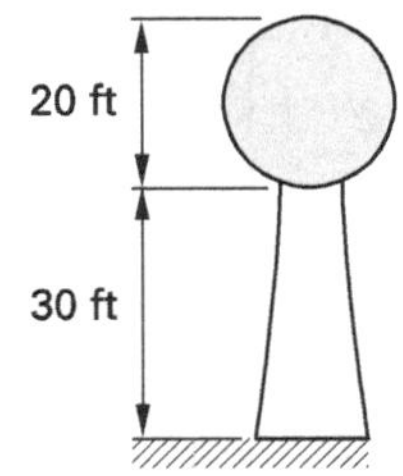

Using ASCE/SEI7, the design wind force on the tank is most nearly

(A) 2500 lbf

(B) 3300 lbf

(C) 3500 lbf

(D) 4900 lbf

Hint: Use ASCE/SEI7 Chap. 29.

PROBLEM 12

The plain concrete foundation wall shown supports a 12 in nominal concrete masonry unit (CMU) wall. The foundation wall bears on soil composed of poorly graded clean sands and is laterally supported top and bottom.

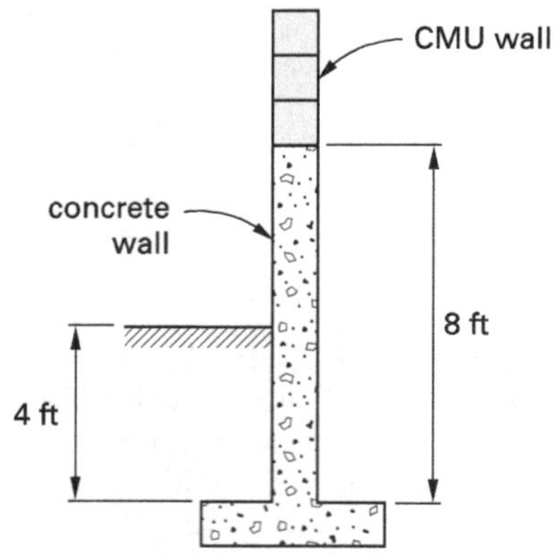

Using the prescriptive criteria found in the *International Building Code* (IBC), the minimum thickness of the foundation wall is most nearly

(A) 7.5 in

(B) 8.0 in

(C) 9.5 in

(D) 12 in

Hint: Use IBC Sec. 1807 to determine the minimum thickness.

PROBLEM 13

The roof of a multistory building is steel composed of steel joists spaced 5 ft on center and spanning 30 ft between steel girders. The girders span 50 ft between columns. The dead load on the flat roof is 32 lbf/ft^2.

Using the IBC, what is most nearly the live load on an interior roof girder?

(A) 12 kips

(B) 18 kips

(C) 26 kips

(D) 30 kips

Hint: Refer to IBC Chap. 16, Structural Loads.

TEMPORARY STRUCTURES AND OTHER TOPICS

PROBLEM 14

A 30 ft tall office building under construction is located 15 ft from its lot line. According to the *International Building Code* (IBC), which safeguard is required to protect pedestrians during construction?

(A) construction railings

(B) barriers

(C) barriers and covered walkway

(D) none

Hint: Refer to IBC Chap. 33.

PROBLEM 15

According to the *International Building Code* (IBC), which of the following statements is true regarding construction documents?

I. The size and location of all structural members must be shown.

II. Information on seismic loads need not be shown if wind governs the design of the lateral force-resisting system.

III. The design wind speed must be shown regardless of whether wind loads govern the design of the lateral force-resisting system.

IV. Live load reductions must be shown.

(A) I only

(B) I and II only

(C) I and III only

(D) I, III, and IV only

Hint: Refer to IBC Sec. 1603.

Codes and Construction

PROBLEM 16

Design of formwork must include consideration of all of the following factors EXCEPT

(A) rate and method of placing concrete

(B) special form requirements

(C) design loads, including vertical, horizontal, and impact loads

(D) construction loads, including vertical, horizontal, and impact loads

Hint: Refer to ACI 318 Chap. 26.11.

PROBLEM 17

A reinforced brick masonry residence located in Los Angeles is designed using strength design provisions. According to *Building Code Requirements and Specification for Masonry Structures* (TMS 402/602), which of the following requirements must be met during design and construction of this building?

I. Verify placement of reinforcement prior to grouting.

II. Verify placement of grout continuously during construction.

III. Observe preparation of mortar specimens.

IV. Verify the compressive strength of masonry prior to construction and every 5000 ft^2 during construction.

(A) I and II only

(B) I and III only

(C) I, II, and III

(D) none of the above

Hint: These requirements are part of a quality assurance program.

PROBLEM 18

Wood formwork consisting of 3×4 joists and stringers is used to support a 4 in normalweight concrete slab. The live load during construction is 50 lbf/ft^2. Given that the maximum spacing of the stringers limited by bending in the joists is $(4466\ lbf/w)^{1/2}$, the maximum spacing of the stringers as limited by bending in the stringers is 523.4 lbf-ft/w, and the maximum spacing of the stringers as limited by the load to the post is 512.8 lbf-ft/w. The maximum spacing of the stringers is most nearly

(A) 5.1 ft

(B) 5.9 ft

(C) 6.7 ft

(D) 9.5 ft

Hint: Determine the load on the stringers.

PROBLEM 19

According to OSHA safety and health regulations for construction, which statement is INCORRECT?

(A) Formwork must be designed to support all loads, both vertical and lateral, that could reasonably be expected to be applied to it.

(B) Drawings or plans for formwork, including shoring equipment, must be available at the jobsite.

(C) Single post shores may be offset by a maximum of 5%.

(D) The erected shoring must be inspected by an engineer qualified in structural design.

Hint: Consult OSHA CFR 29, Part 1926, Subpart Q.

SOLUTION 1

Chapter 8 of ACI 318 covers two-way slab systems. Section 8.10 covers the direct design method. Statement I is true.

Section 8.10.2.1 of ACI 318 requires that there be at least three continuous spans in each direction in order to justify use of the direct design method. Statement II is false.

Section 8.11 of ACI 318 gives the requirements for the equivalent frame method. There are no limitations in this section on the number and length of spans that can be analyzed using this method. Statement III is true.

Section 8.10.2.6 of ACI 318 limits the live load to two times the dead load for slabs using the direct design method. Statement IV is true.

The answer is (C).

Why Other Options Are Wrong

(A) Although statement I is true, statement II is false.

(B) Although statements I and III are true, so is statement IV.

(D) Although statements I, III, and IV are true, statement II is false.

SOLUTION 2

Section 12.14 of ASCE/SEI7 contains the simplified alternative structural design procedure for simple bearing wall buildings. The seismic base shear is

$$V = \left(\frac{FS_{\text{DS}}}{R}\right)W \quad \text{[ASCE/SEI7 Eq. 12.14-12]}$$

The design elastic response acceleration at the short period, S_{DS}, is given as

$$S_{\text{DS}} = \tfrac{2}{3}F_aS_S \quad \text{[ASCE/SEI7 Sec. 12.14.8.1]}$$

ASCE/SEI7 Sec. 12.14.8.1 permits F_a to be taken as 1.4 for soil sites or be determined in accordance with Table 11.4-1 of Sec. 11.4.3. For simplicity, take F_a to be 1.4. The maximum considered earthquake ground motion for 0.2 sec spectral response, S_S, is 0.3 g.

$$\begin{aligned} S_{\text{DS}} &= \tfrac{2}{3}F_aS_S \\ &= \left(\frac{2}{3}\right)(1.4)(0.3) \\ &= 0.28 \end{aligned}$$

$F = 1.0$ for one-story buildings. R is the response modification factor from ASCE/SEI7 Table 12.14-1 and is a function of the basic seismic-force-resisting system. For a bearing wall system, use Part A of the table. Based on System 9, for ordinary reinforced masonry shear walls, R is 2.0.

$$\begin{aligned} V &= \left(\frac{FS_{\text{DS}}}{R}\right)W = \left(\frac{(1.0)(0.28)}{2.0}\right)W \\ &= 0.14W \end{aligned}$$

The answer is (B).

Why Other Options Are Wrong

(A) This incorrect solution finds the response modification factor in ASCE/SEI7 Table 12.14-1 based on System 2, which gives an R of 4 for ordinary reinforced concrete shear walls.

(C) This incorrect solution assumes the design elastic response acceleration at the short period, S_{DS}, to be given as 0.30.

(D) This incorrect solution finds the response modification factor in ASCE/SEI7 Table 12.14-1 based on System 2, which gives an R of 4 for ordinary reinforced concrete shear walls, and assumes the design elastic response acceleration at the short period, S_{DS}, to be given as 0.30.

SOLUTION 3

Section 8.5.4 of ACI 318 covers openings in slab systems. Without special analysis, openings in a two-way slab system are limited by where they occur.

Determine the width of the column strips in each direction. Column strip width on each side of a column centerline is $0.25l_1$ or $0.25l_2$, whichever is less [ACI 318 Sec. 8.4.1.5].

North/South Direction:

$$\begin{aligned} l_1 &= 20 \text{ ft} \\ l_2 &= 30 \text{ ft} \\ \text{column strip width} &= 0.25l_1 = (0.25)(20 \text{ ft}) = 5.0 \text{ ft} \\ \text{column strip width} &= 0.25l_2 = (0.25)(30 \text{ ft}) = 7.5 \text{ ft} \end{aligned}$$

The lesser value controls.

East/West Direction:

$$\begin{aligned} l_1 &= 30 \text{ ft} \\ l_2 &= 20 \text{ ft} \\ \text{column strip width} &= 0.25l_1 = (0.25)(30 \text{ ft}) = 7.5 \text{ ft} \\ \text{column strip width} &= 0.25l_2 = (0.25)(20 \text{ ft}) = 5.0 \text{ ft} \end{aligned}$$

The lesser value controls.

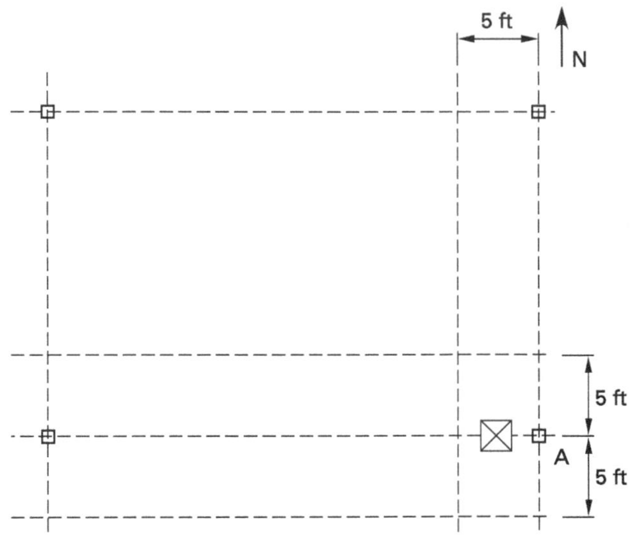

The opening falls within intersecting column strips. ACI 318 Sec. 8.5.4.2(b) limits openings in intersecting column strips to not more than ⅛ the width of the column strip in either span.

$$a = b \leq \left(\frac{1}{8}\right)(10 \text{ ft}) = 1.25 \text{ ft}$$

The maximum opening is

$$ab = (1.25 \text{ ft})^2 = 1.56 \text{ ft}^2 \quad (1.6 \text{ ft}^2)$$

The answer is (C).

Why Other Options Are Wrong

(A) This incorrect solution assumes that since no special analysis is done, no openings are permitted.

(B) In this incorrect solution, one-half the column strip width is used instead of the full column strip width in calculating the opening size.

(D) This incorrect solution does not use the lesser value in calculating the column strip widths according to ACI 318 Sec. 8.4.1.5.

SOLUTION 4

Section 1607.12 of the IBC states that roofs must be designed for the appropriate live loads. The minimum uniformly distributed roof live loads are given in IBC Table 1607.1 and are permitted to be reduced per IBC Eq. 16-26. The minimum roof live load, L_o, is given in IBC Table 1607.1 as 20 lbf/ft^2 for an ordinary pitched roof. The reduced live load is

$$L_r = L_o R_1 R_2 \quad \text{[IBC Eq. 16-26]}$$

The reduction factors are based on the tributary area of the structural member and the slope of the roof. The tributary area of the column is

$$A_t = (20 \text{ ft})(15 \text{ ft}) = 300 \text{ ft}^2$$

The number of inches of rise per foot of the roof is

$$F = 6$$

The reduction factors are calculated using IBC Eq. 16-28 for R_1 if 200 $\text{ft}^2 < A_t <$ 600 ft^2 and using IBC Eq. 16-31 for R_2 if $4 < F < 12$.

$$\begin{aligned} R_1 &= 1.2 - 0.001A_t \\ &= 1.2 - (0.001)(300) \\ &= 0.90 \\ R_2 &= 1.2 - 0.05F \\ &= 1.2 - (0.05)(6) \\ &= 0.90 \end{aligned}$$

The reduced live load is

$$\begin{aligned} L_r &= L_o R_1 R_2 \\ &= \left(20 \ \frac{\text{lbf}}{\text{ft}^2}\right)(0.90)(0.90) \\ &= 16.2 \text{ lbf/ft}^2 \end{aligned}$$

The minimum live load on the column is

$$\begin{aligned} P &= L_r A_t \\ &= \left(16.2 \ \frac{\text{lbf}}{\text{ft}^2}\right)(300 \text{ ft}^2)\left(\frac{1 \text{ kip}}{1000 \text{ lbf}}\right) \\ &= 4.86 \text{ kips} \quad (4.9 \text{ kips}) \end{aligned}$$

The answer is (B).

Why Other Options Are Wrong

(A) This incorrect solution uses the live load reduction factors found in IBC Sec. 1607.10 instead of the roof reduction factors.

(C) This incorrect solution does not reduce the roof load.

(D) This incorrect solution calculates the total load, not the live load.

Codes and Construction

SOLUTION 5

Section 1613 of the IBC refers to ASCE/SEI7 for seismic design, which contains limits on story drift for all seismic design categories. (See ASCE/SEI7 Sec. 12.12.) Therefore, statement C is false.

The answer is (C).

Why Other Options Are Wrong

(A) According to the IBC and ASCE/SEI7, the design lateral seismic force is directly proportional to the building weight. For example, ASCE/SEI7 Sec. 12.8 gives the following equation for seismic base shear.

$$V = C_s W \quad \text{[ASCE/SEI7 Eq. 12.8-1]}$$

In this equation, W is the effective seismic weight, which includes the total dead load. As W increases, so does the total design base shear, V. Statement A is true.

(B) Seismic design category is based on seismic use Group and the design spectral response coefficients.

According to IBC Sec. 1604.5, an apartment building is classified as risk category II.

Use IBC Table 1613.3.5(1) to determine that S_{DS} is between 0.33 g and 0.50 g, knowing the apartment building is classified as seismic design category C and risk category II.

(D) IBC Sec. 1613.3.2 covers site class definitions. The exception in this section states, "When the soil properties are not known in sufficient detail to determine the site class, site class D shall be used..." Statement D is true.

SOLUTION 6

A roof member that does not support a ceiling is limited by Table 1604.3 of the IBC to a total load deflection of $l/120$.

$$\begin{aligned}\Delta &= \frac{l}{120} = \left(\frac{60 \text{ ft}}{120}\right)\left(12 \ \frac{\text{in}}{\text{ft}}\right) \\ &= 6.0 \text{ in} \quad (6 \text{ in})\end{aligned}$$

The answer is (D).

Why Other Options Are Wrong

(A) This incorrect solution does not convert the length of the member to inches in calculating the deflection limit. The units do not work out.

(B) This incorrect solution uses the deflection limit for total load with plastered ceilings, $l/240$, instead of the deflection limit for total load with no ceiling.

(C) This incorrect solution uses the total load deflection limit for nonplastered ceilings, $l/180$, instead of the deflection limit for no ceiling.

SOLUTION 7

The tabulated allowable bending design values must be multiplied by the beam stability factor, C_L, as specified in NDS Sec. 3.3.3. Statement I is false.

The size factor does not apply to structural glulam timber. The adjustments to design values for structural glulam timber are given in NDS Sec. 5.3. Statement II is true.

NDS Sec. 6.3.5 specifies that the referenced design values for round timber piles are based on air-dried conditions. The tabulated design values must be multiplied by the condition treatment factor, C_{ct}. Statement III is false.

The group action factor does not apply to all types of wood connections. Timber rivets and spikes are among those connections to which the group action factor does not apply [NDS Sec. 11.3.1]. Statement IV is false.

The answer is (D).

Why Other Options Are Wrong

(A) This incorrect solution only identifies the first false statement. Statements III and IV are also false.

(B) This incorrect solution identifies the only true statement rather than those that are not true.

(C) This incorrect solution correctly identifies statements I and IV as false but misreads NDS Sec. 6.3.5 to indicate that the treatment factors are already incorporated into the design values.

SOLUTION 8

Section 1608 of the IBC states that design snow loads must be determined in accordance with ASCE/SEI7 Chap. 7. Begin by determining the flat roof snow load, p_f, and adjusting it for the roof slope.

$$p_f = 0.7 C_e C_t I_s p_g \quad \text{[ASCE/SEI7 Eq. 7.3-1]}$$

According to ASCE/SEI7 Table 7-3, the thermal factor, C_t, for ventilated roofs with an R-value over 25 is 1.1.

Then, from the given values, the flat roof snow load is

$$\begin{aligned}p_f &= 0.7 C_e C_t I_s p_g = (0.7)(0.9)(1.1)(1.0)\left(40 \ \frac{\text{lbf}}{\text{ft}^2}\right) \\ &= 27.7 \text{ lbf/ft}^2\end{aligned}$$

To determine the maximum leeward snow load, both balanced and unbalanced loads must be considered. Use ASCE/SEI7 Sec. 7.4 to determine the sloped roof balanced snow load.

$$p_s = C_s p_f$$

The roof slope factor, C_s, is determined in accordance with ASCE/SEI7 Sec. 7.4. It depends on whether the roof is cold or warm, and slippery or non-slippery. An asphalt-shingle roof is considered not slippery. Per ASCE/SEI7 Sec. 7.4.2, cold roofs are those with $C_t > 1.0$, so the roof is cold. Use ASCE/SEI7 Fig. 7-2b to find that C_s is 1.0. The sloped roof balanced snow load is

$$\begin{aligned} p_s &= C_s p_f \\ &= (1.0)\left(27.7 \ \frac{\text{lbf}}{\text{ft}^2}\right) \\ &= 27.7 \ \text{lbf/ft}^2 \end{aligned}$$

As defined by ASCE/SEI7 Sec. 7.6.1, for hip and gable roofs, W is the horizontal eave-to-ridge distance. If W is 20 ft or less, for roofs with simply supported members spanning eave-to-ridge, the unbalanced snow load on the leeward side is equal to Ip_g.

$$Ip_g = (1.0)\left(40 \ \frac{\text{lbf}}{\text{ft}^2}\right) = 40 \ \text{lbf/ft}^2$$

The unbalanced snow load on the windward side is zero.

The unbalanced load condition on the leeward side is greater than the balanced load condition. The maximum snow load on the leeward roof is 40 lbf/ft^2.

The answer is (D).

Why Other Options Are Wrong

(A) This incorrect solution finds the unbalanced roof snow load on the windward side.

(B) This incorrect solution finds the sloped roof balanced snow load for a slippery roof and ignores the unbalanced condition.

(C) This incorrect solution finds the correct sloped roof balanced snow load but ignores the unbalanced condition.

SOLUTION 9

Section 12.8.1 of ASCE/SEI7 defines seismic base shear as directly proportional to effective seismic weight, W. The effective seismic weight is the total dead load plus portions of other loads as listed in ASCE/SEI7. Therefore, as the building dead load increases, so does the seismic base shear. Statement I is true.

ASCE/SEI7 Sec. 12.7.2 states that where the flat snow load exceeds 30 psf, the effective seismic weight shall include 20% of the uniform design snow load. Statement III is also true.

The answer is (B).

Why Other Options Are Wrong

(A) In areas of high seismic activity, buildings perform best when they have a regular floor plan to more uniformly distribute the seismic loads. Irregular plans often have greater building eccentricities and greater differential seismic movements that are harder to accommodate in the building design. Statement II is false.

(C) In the Northridge earthquake and others, buildings with flexible (or soft) first stories and heavy roofs often failed when the flexible first story swayed beyond the design limits and collapsed. In general, it is preferable to have the more flexible story above a stiffer one. Statement IV is false.

(D) Even though statements I and III are true, statement II is false.

SOLUTION 10

ACI 318 Sec. 25.4.2 contains the development length requirements for bars in tension. The development length is measured from the point where the stress in the bars is at maximum. For the positive moment, the distance is measured from the center of the span. Since the clear spacing of the bars is greater than twice the diameter of the bars and the clear cover is greater than the diameter of the bars, the development length for a no. 9 bar is given in ACI 318 Sec. 25.4.2.2 as

$$l_d = \left(\frac{f_y \psi_t \psi_e}{20\lambda\sqrt{f_c'}}\right) d_b \geq 12 \ \text{in}$$

$$d_b = 1.128 \ \text{in}$$

ACI 318 Sec. 25.4.2.4 gives the factors used in this equation.

$$\begin{aligned} \psi_t &= 1.0 \\ \psi_e &= 1.0 \\ \lambda &= 1.0 \\ l_d &= \left(\frac{f_y \psi_t \psi_e}{20\lambda\sqrt{f_c'}}\right) d_b \\ &= \left(\frac{\left(60{,}000 \ \frac{\text{lbf}}{\text{in}^2}\right)(1.0)(1.0)}{(20)(1.0)\sqrt{6000 \ \frac{\text{lbf}}{\text{in}^2}}}\right)(1.128 \ \text{in}) \\ &= 43.7 \ \text{in} \end{aligned}$$

Codes and Construction

ACI 318 Sec. 9.7.3.3 states that reinforcement shall extend beyond the point at which it is no longer needed for a distance equal to the greater of the following.

$$d = h - \text{cover} - \frac{d_b}{2} = 30 \text{ in} - 1.5 \text{ in} - \frac{1.128 \text{ in}}{2} = 27.9 \text{ in}$$

$$12d_b = (12)(1.128 \text{ in}) = 13.5 \text{ in}$$

The point at which the positive moment reinforcement is no longer needed is the point of inflection. Since the point of inflection is 3 ft from the face of the support, the bars must extend to within 36 in − 27.9 in of the face of the support, which equals 8.1 in.

ACI 318 Sec. 9.7.3.8.2 states that at least 1/4 of the positive moment reinforcement must extend into the support at least 6 in. This requirement controls.

The answer is (B).

Why Other Options Are Wrong

(A) This incorrect solution assumes that since there is no positive moment at the support, no reinforcement is required. However, ACI 318 Sec. 9.7.3.8.2 indicates otherwise.

(C) This incorrect solution assumes that the point where the positive moment reinforcement is no longer needed is the face of the support. The point of inflection is actually the point where the positive moment goes to zero.

(D) This incorrect solution assumes the development length to be measured from the face of the support. Development length should be measured from the point where the stress in the bars is at maximum. For the positive moment, the distance should be measured from the center of the span.

SOLUTION 11

ASCE/SEI7 Chap. 29 covers wind loads on other structures such as water tanks. The design wind force for other structures is given in ASCE/SEI7 Sec. 29.5 as

$$F = q_z G C_f A_f \quad \text{[ASCE/SEI7 Eq. 29.5-1]}$$

The velocity pressure, q_z, is given as 25 lbf/ft^2. The gust effect factor, G, is given as 0.85. The projected area, A_f, is given as 315 ft^2. Find the force coefficient, C_f.

The force coefficient for tanks is found in ASCE/SEI7 Fig. 29.5-1 and is based on the ratio of the structure's height-to-diameter ratio and the velocity pressure. In this case,

$$h = 30 \text{ ft} + 20 \text{ ft} = 50 \text{ ft}$$

$$D = 20 \text{ ft}$$

$$\frac{h}{D} = \frac{50 \text{ ft}}{20 \text{ ft}} = 2.5$$

$$D\sqrt{q_z} = (20 \text{ ft})\sqrt{25 \frac{\text{lbf}}{\text{ft}^2}} = 100$$

For a moderately smooth tank, interpolate ASCE/SEI7 Fig. 29.5-1 to determine the force coefficient.

$$C_f = 0.53$$

The design wind force for the tank is

$$F = q_z G C_f A_f = \left(25 \frac{\text{lbf}}{\text{ft}^2}\right)(0.85)(0.53)(315 \text{ ft}^2) = 3548 \text{ lbf} \quad (3500 \text{ lbf})$$

The answer is (C).

Why Other Options Are Wrong

(A) This incorrect solution is the velocity pressure with incorrect units.

(B) This incorrect solution does not interpolate ASCE/SEI7 Fig. 29.5-1 correctly and uses a value of 0.5 for the force coefficient.

(D) This incorrect solution uses a height of 30 ft when calculating the h/D ratio and the resulting force coefficient.

SOLUTION 12

Section 1807.1.6 of the IBC includes prescriptive requirements for concrete and masonry foundation walls.

Section 1807.1.6.1 of the IBC specifies that the minimum thickness of foundation walls must not be less than the thickness of the wall supported. Therefore, the minimum thickness of the foundation wall must be equal to the concrete masonry wall thickness, or 11.63 in.

Verify that 11.63 in is adequate based on the soil lateral load. IBC Sec. 1610 requires that foundation walls that are laterally supported at the top be designed for at-rest pressure. Using IBC Table 1610.1, poorly graded clean sands and sand-gravel mixes have a unified soil classification of SP and an at-rest design lateral soil load of 60 lbf/ft^2-ft.

IBC Table 1807.1.6.2 specifies the minimum thickness for concrete foundation walls. The minimum thickness depends upon the wall height and the height of the unbalanced backfill. The wall height is 8 ft. The height of unbalanced backfill is 8 ft minus 4 ft, which equals

4 ft. Knowing these values and the design lateral soil load of 60 lbf/ft^2-ft, Table 1807.1.6.2 indicates a minimum wall thickness of 7.5 in is adequate.

The answer is (D).

Why Other Options Are Wrong

(A) This incorrect solution overlooks the requirements of IBC Sec. 1807.1.6.1 for minimum thickness equal to the supported wall thickness.

(B) This incorrect solution overlooks the requirements of IBC Sec. 1807.1.6.1 for minimum thickness equal to the supported wall thickness and uses IBC Table 1807.1.6.3(1) for plain masonry walls instead of plain concrete walls.

(C) This incorrect solution miscalculates the height of unbalanced backfill in IBC Table 1807.1.6.2 as 8 ft instead of 4 ft and overlooks the requirements of IBC Sec. 1807.1.6.1 for minimum thickness equal to the supported wall thickness.

SOLUTION 13

Section 1607 of the IBC covers live loads. Minimum roof live loads are determined in accordance with IBC Sec. 1607.12.2 depending on area supported and slope of the roof. According to IBC Table 1607.1, the roof live load, L_o, is 20 lbf/ft^2. The reduced design live load is calculated as

$$\begin{aligned} L_r &= L_o R_1 R_2 \quad \text{[IBC Eq. 16-26]} \\ &= \left(20\ \frac{\text{lbf}}{\text{ft}^2}\right) R_1 R_2 \end{aligned}$$

The area supported by an interior girder is

$$A_t = bl = (50\ \text{ft})(30\ \text{ft}) = 1500\ \text{ft}^2$$

The reduction factor, R_1, for tributary areas 600 ft^2 or greater is 0.6 [IBC Eq. 16-29]. A flat roof has a slope factor, F, of 0. Therefore, the reduction factor, R_2, is 1 [IBC Eq. 16-30].

The reduced roof live load on the girder is

$$L_r = \left(20\ \frac{\text{lbf}}{\text{ft}^2}\right) R_1 R_2 = \left(20\ \frac{\text{lbf}}{\text{ft}^2}\right)(0.6)(1) = 12.0\ \text{lbf/ft}^2$$

The live load on the girder is

$$\begin{aligned} P_L = L_r A_t &= \left(12.0\ \frac{\text{lbf}}{\text{ft}^2}\right)\left(\frac{1\ \text{kip}}{1000\ \text{lbf}}\right)(1500\ \text{ft}^2) \\ &= 18.0\ \text{kips} \quad (18\ \text{kips}) \end{aligned}$$

The answer is (B).

Why Other Options Are Wrong

(A) This incorrect solution applies the alternate floor live load reduction found in IBC Sec. 1607.10 to the roof live load and calculates the reduction incorrectly.

(C) This incorrect solution applies the alternate floor live load reduction found in IBC Sec. 1607.10 to the roof live load and makes a mathematical error in calculating the area supported.

The area supported by an interior girder is incorrectly calculated as

$$A = bl = (50\ \text{ft})(30\ \text{ft}) = 1500\ \text{ft}^2$$

(D) This incorrect solution does not use IBC Eq. 16-26 to calculate the roof live load.

$$P_L = LA = \left(20\ \frac{\text{lbf}}{\text{ft}^2}\right)\left(\frac{1\ \text{kip}}{1000\ \text{lbf}}\right)(1500\ \text{ft}^2) = 30\ \text{kips}$$

SOLUTION 14

IBC Chap. 33 Sec. 3306 covers safeguards during construction and the protection of pedestrians. IBC Table 3306.1 lists the type of safeguards required to protect pedestrians during construction based on a building's height and its distance from construction to the lot line. In this case, the distance from construction to the lot line is 15 ft, which is one-half the height of construction. For buildings over 8 ft in height and a distance from construction to the lot line of over 5 ft but no more than one-half the height of construction, barriers are required.

The answer is (B).

Why Other Options Are Wrong

(A) This incorrect solution ignores that construction railings are only required when the height of construction is 8 ft or less, and the distance from construction to the lot line is less than 5 ft.

(C) This incorrect solution ignores that barriers and covered walkways are only required when the height of construction is more than 8 ft and the distance from construction to the lot line is less than in this case.

(D) This incorrect solution ignores that barriers are required for buildings over 8 ft in height and a distance from construction to the lot line of over 5 ft but no more than one-half the height of construction.

SOLUTION 15

According to IBC Sec. 1603, construction documents must show the size, section, and relative locations of structural members (statement I); floor live loads, including live load reductions (statement IV); and information on seismic and wind loads, regardless of whether they govern the design or not (statement III).

The answer is (D).

Why Other Options Are Wrong

(A) This incorrect solution ignores that while I is true, so are III and IV.

(B) This incorrect solution ignores that II is false. Information on seismic loads must be shown regardless of whether seismic loads govern the design of the lateral force-resisting system.

(C) This incorrect solution ignores that while I and III are true, so is IV.

SOLUTION 16

ACI 318 Sec. 26.11.1.2 specifies that design of formwork must include consideration of rate and method of placing concrete; construction loads, including vertical, horizontal, and impact loads; and special form requirements. Design loads are not considered in the design of formwork.

The answer is (C).

Why Other Options Are Wrong

(A) This incorrect solution ignores that ACI 318 Sec. 26.11.1.2 specifies that design of formwork must include consideration of the rate and method of placing concrete.

(B) This incorrect solution ignores that ACI 318 Sec. 26.11.1.2 specifies that design of formwork must include consideration of special form requirements.

(D) This incorrect solution ignores that ACI 318 Sec. 26.11.1.2 specifies that design of formwork must include consideration of construction (not design) loads, including vertical, horizontal, and impact loads, and special form requirements.

SOLUTION 17

TMS 402 Sec. 3.1 contains the quality assurance requirements for masonry structures. There are three levels of quality assurance, depending on the risk category as assigned by ASCE/SEI7 and the method of design.

A residence is risk category II. Since this building was designed using the strength design provisions, it was *not* designed using TMS 402 Part 4: Prescriptive Design Methods (Chap. 12, Chap. 13, or Chap. 14) or Appendix A, which cover veneer, glass unit masonry, masonry partition walls, and empirical masonry, respectively. TMS 402 Sec. 3.1.2.2 requires such buildings to comply with TMS 402 Table 3.1.2.

TMS 402 Table 3.1.2 (Level B quality assurance) contains, among others, the following inspection requirements.

Prior to grouting, verify

- grout space
- grade and size of reinforcement
- placement of reinforcement
- proportions of site-prepared grout
- construction of mortar joints

Continuously verify that the placement of grout is in compliance.

Observe preparation of grout specimens, mortar specimens, and/or prisms.

Verify compressive strength of masonry, f'_m, prior to construction.

Statements I, II, and III are true for Level B quality assurance. Statement IV is a provision of Level C quality assurance that requires continual verification of f'_m throughout construction (every 5000 ft^2).

The answer is (C).

Why Other Options Are Wrong

(A) This incorrect solution correctly identifies statements I and II as true, but statement III is also true.

(B) This incorrect solution identifies statements I and III as true, but does not recognize statement II as applicable to Level B quality assurance. TMS 402 requires continuous verification of grouting.

(D) This incorrect solution mistakenly applies the provisions for buildings designed using TMS 402 Part 4. Strength design is found in TMS 402 Part. 3.

SOLUTION 18

The load on the stringers is

$$w = D + L$$

The weight of the slab is

$$D = t\gamma = (4\text{ in})\left(\frac{1\text{ ft}}{12\text{ in}}\right)\left(150\ \frac{\text{lbf}}{\text{ft}^3}\right) = 50\text{ lbf/ft}^2$$

The live load during construction is given as

$$L = 50\text{ lbf/ft}^2$$

$$w = D + L = 50\ \frac{\text{lbf}}{\text{ft}^2} + 50\ \frac{\text{lbf}}{\text{ft}^2} = 100\text{ lbf/ft}^2$$

The maximum spacing is the lesser of the spacing limited by the bending of the joist, s_j, the spacing limited by the bending of the stringer, s_s, or the spacing limited by the load to the post, s_p.

$$s_j = \sqrt{\frac{4466 \text{ lbf}}{w}} = \sqrt{\frac{4466 \text{ lbf}}{100 \frac{\text{lbf}}{\text{ft}^2}}}$$
$$= 6.68 \text{ ft}$$

$$s_s = \frac{523.4 \frac{\text{lbf}}{\text{ft}}}{w} = \frac{523.4 \frac{\text{lbf}}{\text{ft}}}{100 \frac{\text{lbf}}{\text{ft}^2}}$$
$$= 5.23 \text{ ft}$$

$$s_p = \frac{512.8 \frac{\text{lbf}}{\text{ft}}}{w} = \frac{512.8 \frac{\text{lbf}}{\text{ft}}}{100 \frac{\text{lbf}}{\text{ft}^2}}$$
$$= 5.13 \text{ ft} \quad (5.1 \text{ ft})$$

The maximum spacing of the stringer is 5.1 ft.

The answer is (A).

Why Other Options Are Wrong

(B) This incorrect solution does not properly convert the slab thickness to feet. A slab thickness of 0.25 ft is used instead of 0.33 ft in calculating the weight of the slab.

(C) This incorrect solution is the largest of the spacings rather than the smallest.

(D) This incorrect solution does not include the live load in the calculation of the load on the stringers, w.

SOLUTION 19

OSHA CFR 29, Part 1926, Subpart Q, specifies the requirements for concrete construction.

OSHA Std. 1926.703(a)(1) states that formwork must be designed, fabricated, erected, supported, braced, and maintained so that it will be capable of supporting without failure all vertical and lateral loads that may reasonably be anticipated to be applied to the formwork.

OSHA Std. 1926.703(a)(2) states that drawings or plans, including all revisions, for the jack layout, formwork (including shoring equipment), working decks, and scaffolds must be available at the jobsite.

OSHA Std. 1926.703(b)(8)(ii) states that single post shores must be vertically aligned.

OSHA Std. 1926.703(b)(8)(i) states that the design of the shoring must be prepared by a qualified designer and the erected shoring must be inspected by an engineer qualified in structural design.

The answer is (C).

Why Other Options Are Wrong

(A) This incorrect solution is true per OSHA regulations.

(B) This incorrect solution is true per OSHA regulations.

(D) This incorrect solution is true per OSHA regulations.

3 Design and Details of Structures

COMPONENT DESIGN AND DETAILING

PROBLEM 1

A 20 in deep normal weight concrete cantilever beam is reinforced with four no. 7 uncoated bars at the top of the beam. Half of the bars are spliced at each splice location using a lap splice. Based on the moment at the splice location, only two no. 7 bars are required at that location. The beam has the following properties.

specified compressive strength of concrete	3000 lbf/in^2
yield stress of reinforcement	60,000 lbf/in^2
clear cover of reinforcement	2 in
clear spacing of reinforcement	4 in

The splice length of the reinforcing bars is most nearly

(A) 42 in

(B) 63 in

(C) 81 in

(D) 94 in

Hint: Refer to ACI Sec. 25.5.2 for lap splice requirements.

PROBLEM 2

The base plate of a W12 × 72 ASTM A992 structural steel column bears directly on an 8 ft × 8 ft concrete spread-footing. The base plate is made of ASTM A36 steel and is limited in size to 14 in × 16 in. The column supports a factored load of 630 kips.

compressive strength of concrete	3.0 kips/in^2
yield stress of steel	36 kips/in^2

Using LRFD and the AISC *Steel Construction Manual*, the thickness of the base plate is most nearly

(A) 0.915 in

(B) 0.994 in

(C) 1.26 in

(D) 1.34 in

Hint: Start by determining the bearing stress on the concrete.

PROBLEM 3

A steel column supports an unfactored concentric dead load of 130 kips and an unfactored concentric live load of 390 kips. The effective length with respect to the major axis is 32 ft. The effective length with respect to the minor axis is 18 ft. Using LRFD, what is the lightest ASTM A992 W shape that can be used for the column if the column depth cannot exceed 12 in (nominal)?

(A) W12 × 87

(B) W12 × 136

(C) W12 × 170

(D) W12 × 252

Hint: Use the columns tables in Part 4 (Column Design) of the AISC *Steel Construction Manual*.

PROBLEM 4

A 6.5 ft × 6.5 ft concrete spread footing supports a centrally located 12 in × 12 in concrete column. The dead load on the column, including the column weight, is 50 kips. The live load is 75 kips. The soil report indicates an allowable soil pressure of 4000 lbf/ft^2. Disregard the weight of the soil. The critical (plan) area, b_1b_2, for punching shear is 150 in^2. The ultimate punching shear is most nearly

(A) 5 kips

(B) 120 kips

(C) 130 kips

(D) 180 kips

Hint: Calculate the ultimate load on the footing.

PROBLEM 5

A rectangular combined footing is used near a property line to support two 12 in square concrete columns as shown. The allowable soil pressure is 2000 lbf/ft^2. The compressive strength of concrete is 3000 lbf/in^2.

column 1	
dead load	30 kips
live load	60 kips
moment due to dead load	30 ft-kips
moment due to live load	40 ft-kips

column 2	
dead load	60 kips
live load	60 kips
moment due to dead load	30 ft-kips
moment due to live load	40 ft-kips

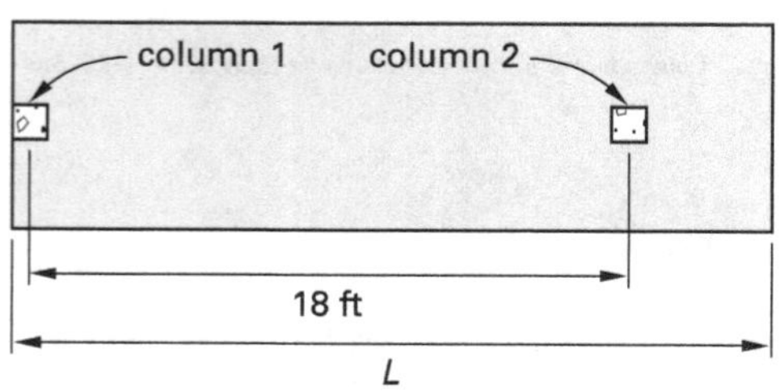

plan

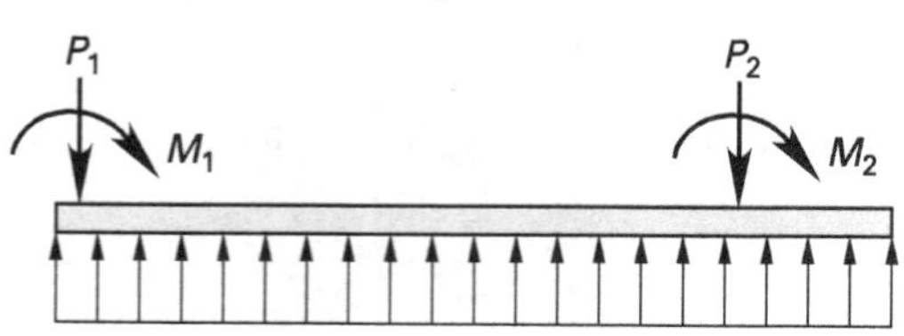

free-body diagram

What is most nearly the minimum footing length, L, that will create uniform soil pressure?

(A) 21.6 ft

(B) 21.8 ft

(C) 22.6 ft

(D) 23.0 ft

Hint: For the soil pressure to be uniform, the resultant, R, must be at the centroid of the footing area.

PROBLEM 6

A 12 in unreinforced concrete masonry wall supports joists spaced 12 in on center. The reaction from each joist is 700 lbf. The wall is grouted solid.

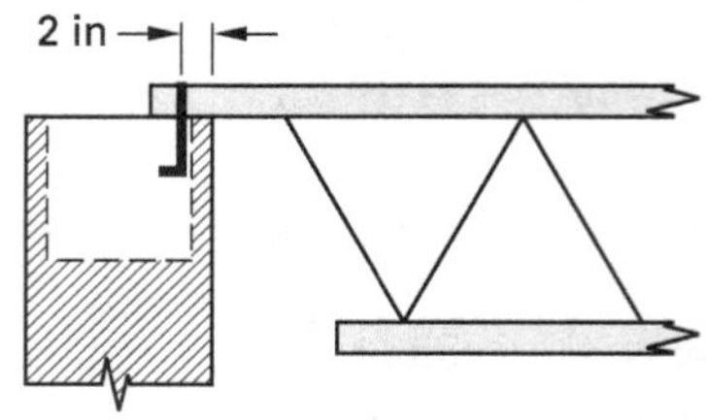

The stress on the wall due to the joists is most nearly

(A) 5.0 lbf/in^2 (compression)

(B) 5.0 lbf/in^2 (tension)

(C) 5.0 lbf/in^2 (compression), 10 lbf/in^2 (tension)

(D) 15 lbf/in^2 (compression), 5.0 lbf/in^2 (tension)

Hint: Determine the eccentricity of the load.

PROBLEM 7

A simply supported reinforced concrete beam carries a uniform dead load of 800 lbf/ft and a uniform live load of 1200 lbf/ft. The beam span length is 24 ft. The beam is constructed of normalweight concrete with a compressive strength of 6000 lbf/in^2.

beam width	16 in
beam height	30 in
beam depth to reinforcement	24 in
area of reinforcing steel	7.9 in^2
moment of inertia of cracked section	15,500 in^4
cracking moment	110 ft-kips

The maximum immediate deflection is most nearly

(A) 0.12 in

(B) 0.17 in

(C) 0.21 in

(D) 0.26 in

Hint: Immediate deflection does not include creep and shrinkage effects.

PROBLEM 8

A truss made from ASTM A36 steel is used to carry vehicular traffic. The diagonal compression members are made from two L6 × 6 × 1/2 in angles with a 3/8 in gusset plate between them as shown. The controlling slenderness ratio, Kl/r, is 155, and the effective length, Kl_x, is 24 ft.

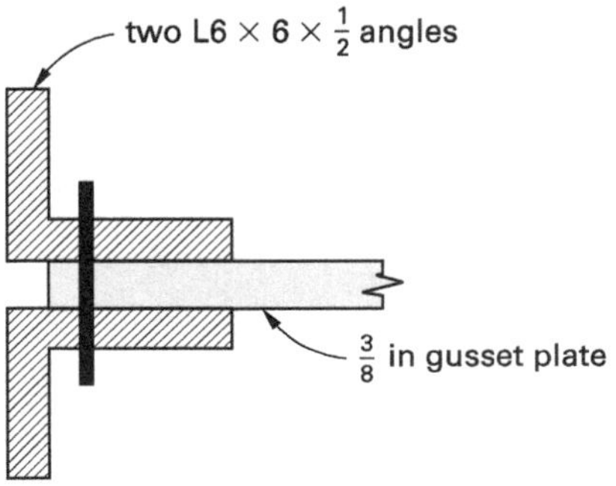

(not to scale)

If the axial load is concentric, what is most nearly the critical compressive stress?

(A) 0.01 kips/in^2

(B) 9 kips/in^2

(C) 10 kips/in^2

(D) 11 kips/in^2

Hint: The legs of the angle are unstiffened elements.

PROBLEM 9

The elevation of an interior girder of a composite two-lane, two-span continuous highway bridge is shown. The web and bearing stiffeners form column sections.

steel specification	ASTM A709
minimum web yield stress	50 kips/in^2
minimum stiffener yield stress	50 kips/in^2
steel modulus of elasticity	29,000 kips/in^2

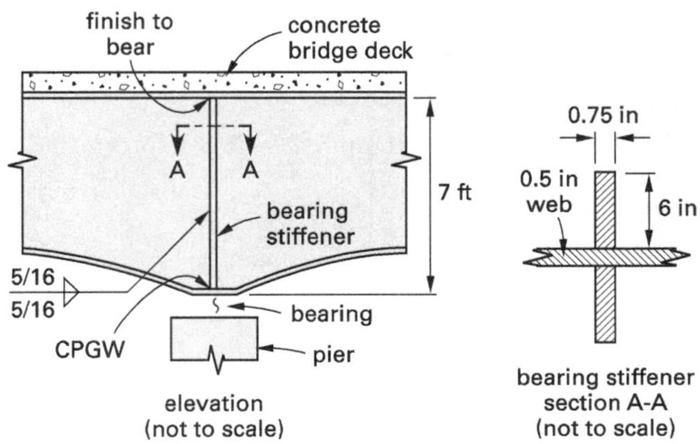

The factored axial resistance of the effective bearing stiffener column section is most nearly

(A) 570 kips

(B) 590 kips

(C) 650 kips

(D) 730 kips

Hint: Use *AASHTO LRFD Bridge Design Specifications* Sec. 6.

PROBLEM 10

Using LRFD, and assuming the compression flange is braced, the maximum flexural strength in a composite W21 × 55 ASTM A992 steel beam with shear connectors constructed without shoring is most nearly limited to

(A) 310 ft-kips

(B) 430 ft-kips

(C) 470 ft-kips

(D) 480 ft-kips

Hint: Use Chap. I of the AISC *Steel Construction Manual.*

PROBLEM 11

The maximum factored shear in the 6 in × 9 in concrete beam shown is 2000 lbf. Assume normalweight concrete with a compressive strength of 3000 lbf/in^2, and the yield stress of the steel reinforcement is 60,000 lbf/in^2.

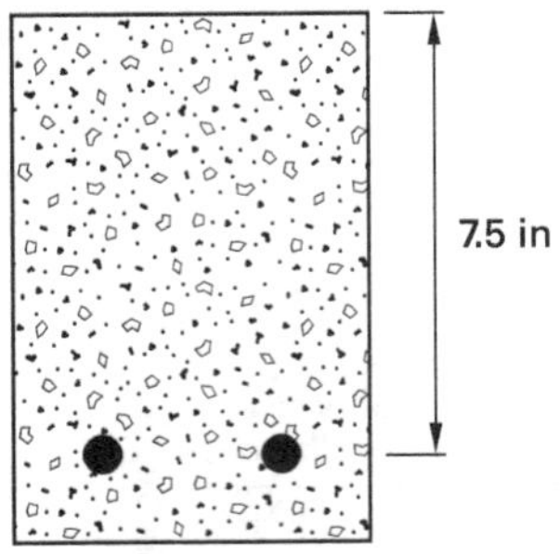

What is most nearly the required shear reinforcement?

(A) no. 3 U-stirrups at 4.0 in spacing

(B) no. 3 U-stirrups at 4.5 in spacing

(C) no. 3 U-stirrups at 24 in spacing

(D) none

Hint: Refer to ACI 318 Sec. 22.5.

PROBLEM 12

A 7.5 in wide × 16 in deep reinforced concrete masonry lintel carries a total uniform load of 1120 lbf/ft, including the self weight of the lintel. The load due to the weight of the wall above the lintel is as shown for the lintel span length, L. The specified compressive strength of the masonry is 3000 lbf/in^2. Two no. 4, grade 60 reinforcing bars are used side-by-side at the bottom of the beam.

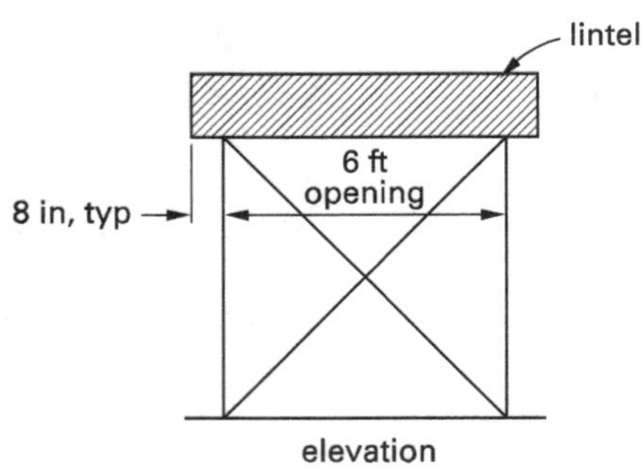

elevation

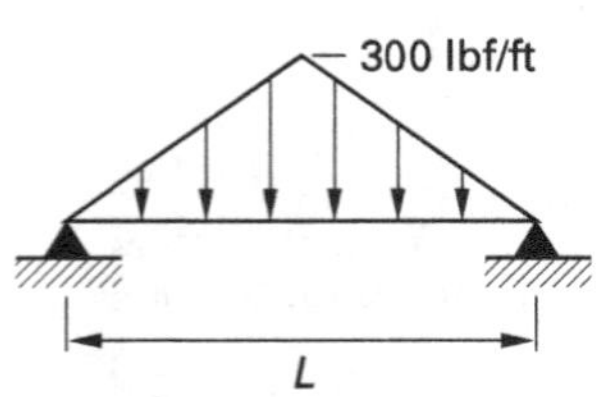

load diagram for weight of wall above lintel

Using ASD, if the maximum moment on the lintel is 88,000 in-lbf what is most nearly the tensile stress in the steel?

(A) 14,000 lbf/in^2

(B) 15,000 lbf/in^2

(C) 17,000 lbf/in^2

(D) 32,000 lbf/in^2

Hint: Calculate the effective depth of the reinforcement first.

PROBLEM 13

The tension member shown is constructed with two 2 × 6 pieces of red oak. Four rows of three ½ in diameter bolts are used to connect the pieces. The spacing between the rows is 0.75 in. The spacing between the closest fasteners, measured parallel to the rows, is 6.5 in.

modulus of elasticity	1.2×10^6 lbf/in^2
load duration factor	1.0
wet service factor	1.0
temperature factor	1.0
geometry factor	1.0
end grain factor	1.0

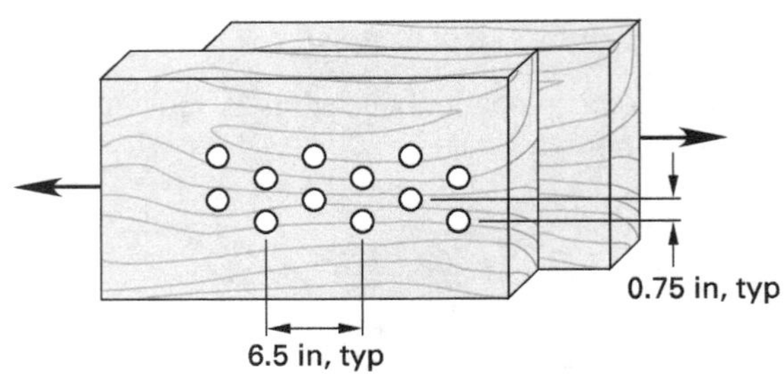

Using ASD, what is most nearly the capacity of the connection?

(A) 610 lbf

(B) 5800 lbf

(C) 7300 lbf

(D) 7700 lbf

Hint: Refer to NDS Chap. 11.

PROBLEM 14

A 10 ft high, 20 ft long special reinforced masonry shear wall is subjected to in-plane seismic loading. The wall is constructed from 12 in concrete masonry units laid in running bond and type S mortar. All reinforcement is adequate to resist the applied shear loads. Which shear reinforcement most nearly complies with the minimum seismic reinforcement requirements?

(A) one no. 5 bar in horizontal bond beams, spaced 24 in on center

(B) one no. 5 bar in horizontal bond beams, spaced 24 in on center, and no. 5 vertical bars, spaced at 24 in on center

(C) two no. 4 bars in horizontal bond beams, spaced 48 in on center, and no. 5 vertical bars, spaced at 24 in on center

(D) two no. 5 bars in horizontal bond beams, spaced 24 in on center, and no. 6 vertical bars, spaced 16 in on center

Hint: Use TMS 402 Sec. 7.3.2.6.

PROBLEM 15

A W14 × 109 steel column carries an axial load of 320 kips and a moment of 200 ft-kips about the y-axis. The centerline of the anchor bolts is 1.5 in outside the column flanges as shown.

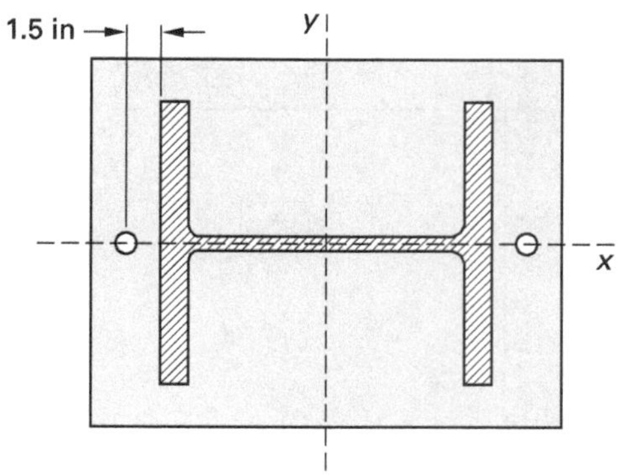

If A307 anchor bolts are used, what size bolt is needed using ASD?

(A) $^5/_8$ in diameter

(B) $^3/_4$ in diameter

(C) $^7/_8$ in diameter

(D) 1 in diameter

Hint: Columns with relatively large moments may be subject to uplift on the base plate.

PROBLEM 16

A reinforced concrete building has a 6 in flat-plate slab on each floor and 12 in diameter columns. The floor-to-floor distance is 13 ft. A column has a factored axial load of 300 kips and equal factored end moments of 100 ft-kips and is subjected to double curvature. The modulus of elasticity of concrete is 3.6×10^6 lbf/in^2 and $kl_u/r = 50$. The column does not have any transverse loads and is not subject to sway. The critical buckling load is most nearly

(A) 400 kips

(B) 680 kips

(C) 1600 kips

(D) 3600 kips

Hint: Use ACI 318 Sec. 6.6.4.4 to determine the critical buckling load.

PROBLEM 17

A reinforced concrete building has a 6 in flat-plate slab on each floor and 12 in diameter columns. The floor-to-floor distance is 13 ft. A column has a factored axial load of 300 kips and equal factored end moments of 100 ft-kips and is subjected to double curvature. The modulus of elasticity of concrete is 3.6×10^6 lbf/in^2 and $kl_u/r = 50$. The critical buckling load is 405 kips. If the column does not have any transverse loads and is not subject to sway, the design moment is most nearly

(A) −1600 ft-kips

(B) 1600 ft-kips

(C) 8100 ft-kips

(D) 13,000 ft-kips

Hint: Use magnified moments to determine the design moment.

PROBLEM 18

A six-story reinforced concrete building is supported on columns placed on a grid and spaced 20 ft apart in each direction. The story height, measured from the top of one slab to the top of the slab above, is 13 ft. At the first level, the distance from the tops of the footings to the top of the first story slab is 18 ft. The columns are 18 in × 18 in and have a moment of inertia of 8800 in^4. Beams span between columns in each direction. Their moment of inertia is 10,000 in^4. The slabs are all 6 in thick two-way spans. 6000 lbf/in^2 concrete with a modulus of elasticity of 4.4×10^6 lbf/in^2 is used throughout the structure. Ignoring the stiffness of the slab, what is most nearly the effective length in the east/west direction for an interior column in an unbraced frame at the second story?

(A) 8.5 ft

(B) 18 ft

(C) 21 ft

(D) 23 ft

Hint: Determine the relative stiffness parameter for the column.

PROBLEM 19

A brick masonry column (nominal area 20 in by 20 in; actual dimensions as shown) is vertically reinforced with four no. 8 grade 60 bars. The masonry has a specified compressive strength of 3500 lbf/in^2. The column's effective height to radius of gyration, h/r, ratio is 72.

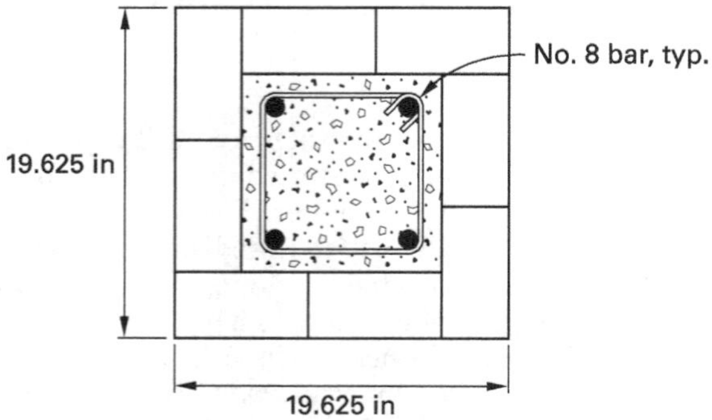

Using allowable stress design, the allowable axial compressive force is most nearly

(A) 284 kips

(B) 294 kips

(C) 306 kips

(D) 309 kips

Hint: Refer to TMS 402 Sec. 8.3.4.

PROBLEM 20

A square concrete footing with a compressive strength of 3000 lbf/in^2 supports an interior round column that is located in the center of the footing. The diameter of the column is 12 in and the footing is 10 ft × 10 ft. The footing thickness is 24 in and punching shear controls. No. 7 bars are spaced equally in each direction at the bottom of the footing. If the shear strength of the reinforcement is ignored, the design two-way shear capacity of the footing is most nearly

(A) 230 kips

(B) 330 kips

(C) 360 kips

(D) 450 kips

Hint: Refer to ACI 318 Sec. 22.6.

PROBLEM 21

A deck is built on the back of a house in a cool humid climate using the design requirements contained in the NDS. A 2 × 8 beam is screwed to the face of a 4 × 4 post as shown. The wood is southern pine. The screws shall have a penetration greater than or equal to ten times the shank diameter. The dead-load plus live-load end reaction of the beam is 605 lbf.

If five 12-gage wood screws are used, what is most nearly the allowable load on the connection?

(A) 450 lbf

(B) 510 lbf

(C) 570 lbf

(D) 800 lbf

Hint: Refer to NDS Chap. 12 for wood screw design values.

PROBLEM 22

The reinforced concrete shear wall shown is made from normalweight concrete with a compressive strength of 4000 lbf/in^2.

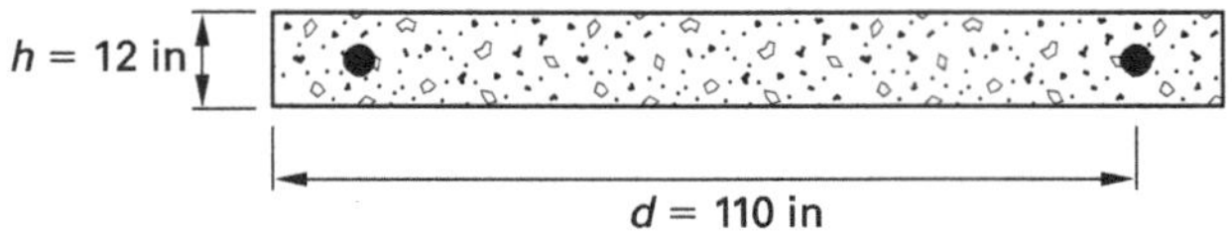

plan
(not to scale)

The nominal shear strength provided by the shear reinforcement is 700 kips. The factored axial gravity load on the wall is 200 lbf/ft. Seismic loads can be ignored. The nominal shear strength of the wall is most nearly

(A) 170 kips

(B) 650 kips

(C) 830 kips

(D) 870 kips

Hint: Refer to ACI 318 Sec. 11.5.

PROBLEM 23

A single-story concrete loadbearing wall supported at the roof and foundation has the following properties.

wall thickness	12 in
concrete compressive strength	4000 lbf/in^2
effective length factor	1.0
length of wall	16 ft
uniform factored axial load	5 kips/ft

Ignore the self-weight of the wall. If the wall is designed using the simplified method, its maximum height is most nearly

(A) 25.0 ft

(B) 26.9 ft

(C) 31.6 ft

(D) 32.0 ft

Hint: Refer to ACI 318 Sec. 11.5.3.

PROBLEM 24

An HP12 × 63 steel pile is driven 100 ft into saturated soft clay. The cohesion, c, is 400 lbf/ft^2, and the adhesion, c_A, is 360 lbf/ft^2. What is most nearly the allowable bearing capacity of the pile if the factor of safety is 3?

(A) 1.7 kips

(B) 49 kips

(C) 69 kips

(D) 150 kips

Hint: The allowable bearing capacity is the sum of the point-bearing capacity and the skin-friction capacity divided by the factor of safety.

PROBLEM 25

A normalweight (0.150 kip/ft^3) reinforced concrete retaining wall is designed to support the loads shown. The soil behind the retaining wall is soil 1. The soil in front of the retaining wall is soil 2.

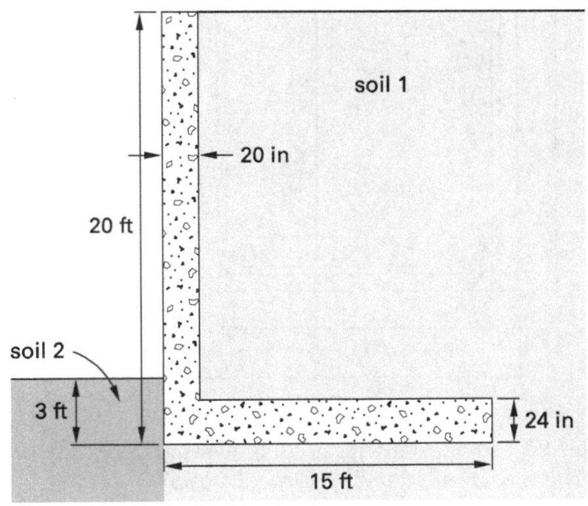

characteristic	soil 1	soil 2
unit weight (kip/ft^3)	0.110	0.100
angle of internal friction (degrees)	28	15
cohesion (kip/ft^2)	0	0.300
active earth pressure coefficient	0.361	0.589
passive earth pressure coefficient	2.77	1.70

The factor of safety against overturning is

(A) 4.16

(B) 4.23

(C) 4.86

(D) 5.91

Hint: The factor of safety against overturning is the ratio of the resisting moment to the overturning moment.

PROBLEM 26

The reinforced concrete beam shown measures 35 in by 40 in and is made from concrete with a compressive strength of 4000 lbf/in^2. Based on its initial design, the beam requires torsional stirrups of 0.05 in^2/in, and no minimum shear reinforcement is required.

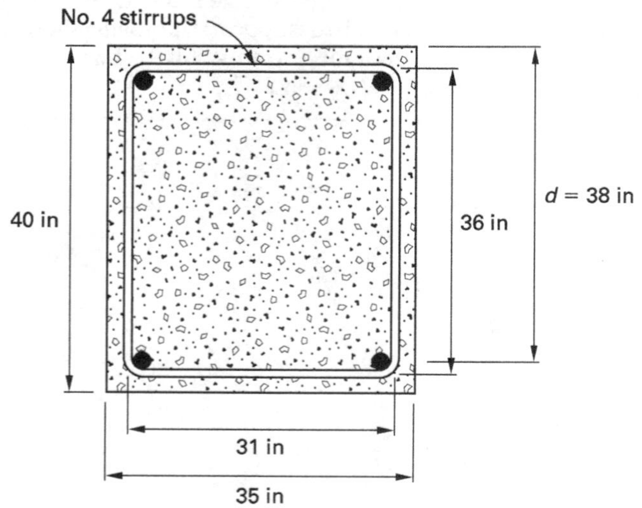

Using no. 4 stirrups with a yield strength of 40,000 lbf/in^2, the maximum spacing is most nearly

(A) 4.0 in

(B) 6.0 in

(C) 12 in

(D) 16 in

Hint: Refer to ACI Sec. 9.7.6 for transverse reinforcement of beams.

PROBLEM 27

Which of the following statements is/are true regarding the design shear strength of a fillet weld?

I. It is limited by the effective throat thickness of the weld.

II. It is directly proportional to the length of the weld.

III. It is determined by the direction of the weld.

IV. It is limited by the nominal strength of the weld metal.

(A) III only

(B) I and II only

(C) I, II, and III

(D) I, II, and IV

Hint: Refer to Sec. J2.2–J2.5 of the AISC *Steel Construction Manual*.

PROBLEM 28

A circular spiral concrete column supports a 300 kip dead load and a 350 kip live load. The concrete compressive strength is 4000 lbf/in^2, and the yield stress of the reinforcement is 60,000 lbf/in^2. If the maximum reinforcement is used, the cross-sectional area of the column is most nearly

(A) 140 in^2

(B) 180 in^2

(C) 200 in^2

(D) 210 in^2

Hint: Refer to ACI 318 Sec. 22.4.2.1.

PROBLEM 29

An ASTM A992 W18 × 40 composite beam with a 4 in concrete slab supports a live load moment of 140 ft-kips and a dead load moment of 80 ft-kips. The dead load includes the weight of the slab and beam. The construction is unshored.

section modulus of the beam	68.4 in^3
transformed section modulus, measured to the bottom of the section	103 in^3
transformed section modulus, measured to the top of the concrete	350 in^3
transformed section modulus, measured to the steel/concrete interface	1680 in^3

Using ASD, the bending stress in the bottom fibers of the steel beam due to dead load is most nearly

(A) 9.3 kips/in^2

(B) 14 kips/in^2

(C) 39 kips/in^2

(D) 1200 lbf/in^2

Hint: In unshored construction, the beam carries the full dead load.

PROBLEM 30

A steel channel strut is connected with high-strength bolts with end-bearing connections. The strut is subjected to a service dead load tensile force of 20 kips. The service live load varies from a 10 kip compressive force to a 50 kip tensile force. It is estimated that the live load may be applied 200 times per day for the life of the structure. The structure is expected to last at least 25 years. The structure is subject to normal atmospheric conditions with temperatures less than 300°F. If the channel is ASTM A36 steel, what is the lightest section that can carry the load?

(A) C7 × 9.8

(B) C6 × 13

(C) C8 × 11.5

(D) C8 × 18.75

Hint: Reversal of the live load must be considered in the design.

PROBLEM 31

A simply supported composite ASTM A992 steel beam has the following properties.

area of concrete	288 in^2
weight of concrete	110 lbf/ft^3
area of steel	11.8 in^2
compressive strength of concrete	3000 lbf/in^2
yield stress of steel	50 kips/in^2

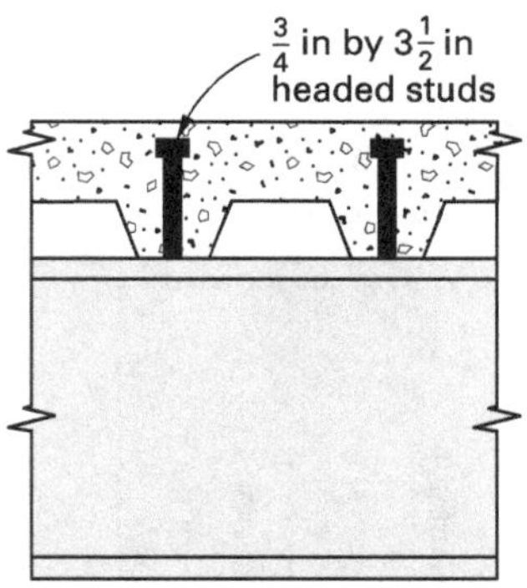

The composite deck has a nominal rib height of 2 in, an average rib width of 3 in, and one $^3/_4$ in diameter stud per rib in the strong position. The total number of studs needed for full composite action is most nearly

(A) 35 studs

(B) 58 studs

(C) 70 studs

(D) 86 studs

Hint: Use Chap. I of the AISC *Steel Construction Manual*.

PROBLEM 32

Southern pine glued laminated timbers (glulams) are used for rafters in a church. The rafters are uniformly loaded and have full lateral support along the compression edges. The tabulated design bending value for the tension zone stressed in tension, F_{bxx}, is 2000 lbf/in^2. The glulams have a width of $8^1/_2$ in and a depth of $27^1/_2$ in and are not tapered. The distance between points of zero moment for the timbers is 15 ft. The loads on each rafter are

dead load moment	60,000 ft-lbf
live load moment	90,000 ft-lbf
wind load moment	100,000 ft-lbf [downward pressure]

The allowable bending design value considering dead plus live load combination is most nearly

(A) 1500 lbf/in^2

(B) 1800 lbf/in^2

(C) 1900 lbf/in^2

(D) 2000 lbf/in^2

Hint: Glulam timber is covered in Chap. 5 of NDS.

PROBLEM 33

An HSS4 × 4 × 3/8 tube beam is welded to the flange of a W10 × 33 ASTM A36 steel column as shown. The reaction at the column is 12,000 lbf. The welds are fillet welds made using the shielded metal arc welding (SMAW) process with E70XX electrodes.

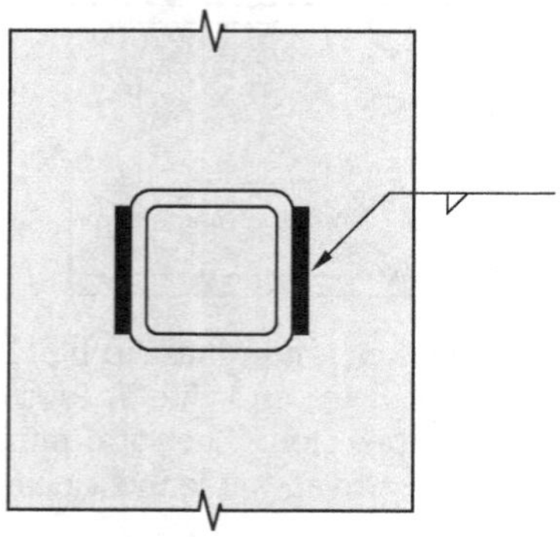

What is the required size of the weld?

(A) 1/8 in

(B) 3/16 in

(C) 1/4 in

(D) The shear exceeds the capacity of a fillet weld.

Hint: Use Sec. J2 of the AISC *Steel Construction Manual*.

PROBLEM 34

A W14 × 22 ASTM A992 steel beam is bolted to a column flange with L3½ × 3½ × 5/16 double angles as shown. A single row of 3/4 in diameter ASTM A325-N bolts is used. Standard size holes are used. Assume that the column-to-clip angle connection is satisfactory.

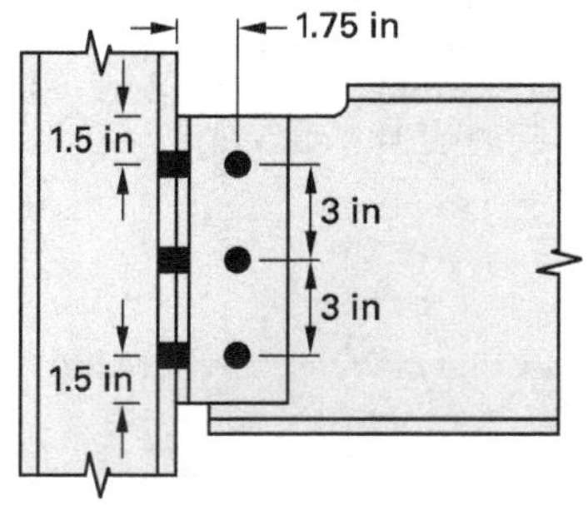

(not to scale)

Using LRFD and the AISC *Steel Construction Manual*, the maximum beam reaction is most nearly

(A) 32 kips

(B) 45 kips

(C) 48 kips

(D) 95 kips

Hint: Refer to Part 10 of the AISC *Steel Construction Manual* for connection design.

PROBLEM 35

An L4 × 8 × ½ angle (LLV) is welded to a W12 × 50 column using the shielded metal arc welding process and E70XX electrodes.

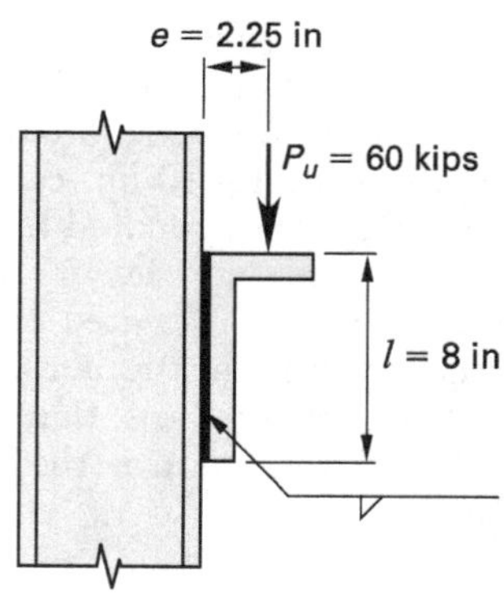

Using LRFD, what is the required fillet weld size if the angle is welded on both sides of the vertical leg only?

(A) 1/8 in

(B) 3/16 in

(C) 1/4 in

(D) 5/16 in

Hint: The weld is subject to both shear and bending.

PROBLEM 36

Plywood sheathing is attached to 2 × 12 roof rafters with 6d box nails. The nails penetrate 1 in into the rafters. The rafters are spaced 16 in on center. All members are southern pine with a modulus of elasticity of 1,700,000 lbf/in^2. Sustained temperatures do not exceed 100°F. If the design uplift wind load on the roof is 36 lbf/ft^2, the maximum nail spacing is most nearly

(A) 7.8 in

(B) 11 in

(C) 12 in

(D) 14 in

Hint: Refer to NDS Chap. 12.

PROBLEM 37

A 20 in × 24 in short concrete column carries an axial factored load of 750 kips and a factored moment of 600 ft-kips. Bars are located on the short faces of the column, and γ is 0.75. The concrete has a compressive strength of 4000 lbf/in^2, and the reinforcement has a yield stress of 60,000 lbf/in^2. The required area of steel is most nearly

(A) 12 in^2

(B) 14 in^2

(C) 19 in^2

(D) 36 in^2

Hint: Use an interaction diagram to determine the required area of steel.

PROBLEM 38

A three-story reinforced concrete building is supported on columns placed on a grid and spaced 20 ft apart in each direction. The first story column length is 18 ft. The second and third story column lengths are 13 ft. Beams span between columns in both directions. 6000 lbf/in^2 concrete with a modulus of elasticity of 4.4 × 10^6 lbf/in^2 is used throughout the structure.

If the gross moment of inertia of the spandrel beams is 10,000 in^4, and the gross moment of inertia of the columns is 8748 in^4, what is the relative stiffness parameter, Ψ, at the top of the first-floor corner column in either direction?

(A) 2.3

(B) 4.6

(C) 5.4

(D) 6.6

Hint: Refer to ACI 318 Chap. 6.

PROBLEM 39

The plan and elevation views of a reinforced concrete bridge pier is shown.

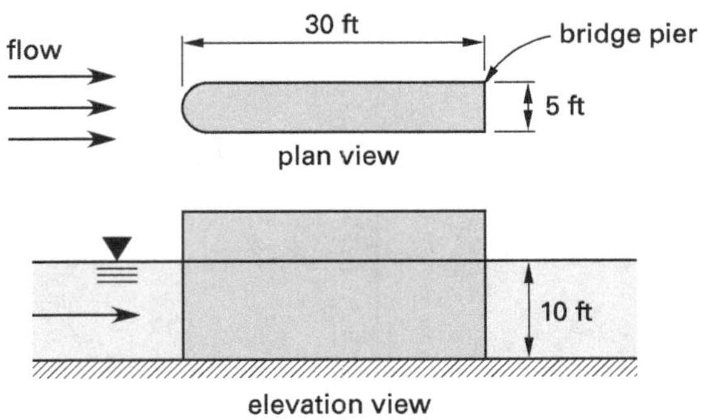

If the water velocity of the stream's design flood is 10 ft/sec, the unfactored longitudinal drag force acting on the upstream edge of the bridge pier is most nearly

(A) 0.35 kips

(B) 3.5 kips

(C) 7.0 kips

(D) 21 kips

Hint: Use *AASHTO LRFD Bridge Design Specifications* Sec. 3.7.

PROBLEM 40

A 4.5 ft × 20 ft rectangular concrete footing supports two 12 in concrete columns.

column 1	
dead load	116 kips
live load	64 kips
column 2	
dead load	70 kips
live load	32 kips

yield stress of reinforcement	60,000 lbf/in^2
compressive strength of concrete	3000 lbf/in^2
depth to reinforcement	15 in

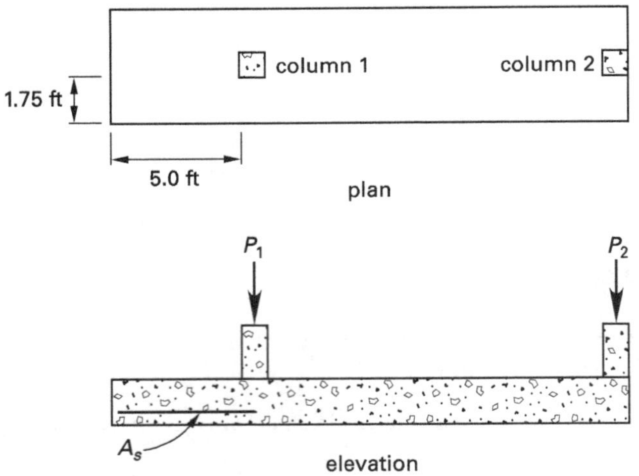

How many no. 6 bars are required in the longitudinal direction between column 1 and the left edge of the footing if the critical design moment is located at the outer face of column 1 and soil pressure is assumed to be uniform?

(A) two

(B) six

(C) seven

(D) nine

Hint: Concrete footing design is based on factored loads.

PROBLEM 41

A concrete masonry non-loadbearing wall is constructed using 12 in units and running bond. The wall is subject to out-of-plane wind loads and is vertically reinforced with grade 60 no. 5 bars located in the center of the wall spaced at 24 in on center. The compressive strength of masonry is 2000 lbf/in^2. Using strength design, the moment capacity of the wall is most nearly

(A) 4300 ft-lbf

(B) 7800 ft-lbf

(C) 8700 ft-lbf

(D) 13,000 ft-lbf

Hint: See TMS 402 Sec. 9.3.5.

PROBLEM 42

A beam-column made from ASTM A992 steel supports a factored axial load of 600 kips at an eccentricity of 12 in about the strong axis. The effective length, KL, is 28 ft. Using LRFD, select a W14 shape to carry the load. Second-order effects need not be considered.

(A) W14 × 176

(B) W14 × 193

(C) W14 × 257

(D) W14 × 311

Hint: Use the equivalent axial load procedure found in Part 4 and Part 6 of the AISC *Steel Construction Manual.*

PROBLEM 43

A 30 ft high, 10 in thick reinforced normalweight concrete shear wall has the following properties.

compressive strength of concrete	6000 lbf/in^2
yield stress of reinforcement	60,000 lbf/in^2
depth to reinforcement	110 in
horizontal length of wall	138 in

The factored in-plane shear force of 60 kips is due to wind loads only. If a single mat of reinforcement is used, the horizontal shear reinforcement required is most nearly

(A) no. 3 at 12 in

(B) no. 4 at 12 in

(C) no. 5 at 12 in

(D) no. 5 at 18 in

Hint: Refer to ACI 318 Chap. 11

PROBLEM 44

A 10 ft high, 30 ft long reinforced concrete shear wall has a compressive strength of 4000 lbf/in^2 and a steel reinforcement yield stress of 60,000 lbf/in^2. The ratio of horizontal shear reinforcement area to gross concrete area of vertical section, ρ_t, is 0.0040. The ratio of vertical shear reinforcement area to gross concrete area of horizontal section, ρ_l, should be at least

(A) 0.0021

(B) 0.0025

(C) 0.0040

(D) 0.0041

Hint: Refer to ACI 318 Sec. 11.6.2.

PROBLEM 45

The cantilevered sheet piling shown has a concentrated lateral load, H, of 10 kips spaced 2 ft on center, horizontally along the top of the piling. The soil specific weight is 110 lbf/ft^3, and the angle of internal friction of the soil is 30°.

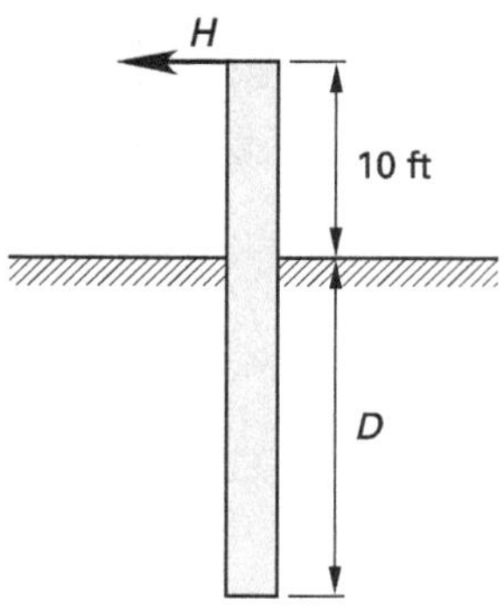

If the factor of safety for the minimum depth is 1.3, the total length of the sheet pile is most nearly

(A) 21 ft

(B) 26 ft

(C) 31 ft

(D) 38 ft

Hint: The solution for this problem is an iterative one.

PROBLEM 46

A wooden pier is supported on kiln-dried round red pine timber piles that extend 20 ft below the water surface into clay soil. A group of three piles connected by a concrete cap carries a 300 kip dead load and a 400 kip live load. The piles have an area of 230 in^2. The column stability factor is 0.62. Using ASD, what is the total adjustment factor for the critical compression load case?

(A) 0.55

(B) 0.56

(C) 0.61

(D) 0.68

Hint: Refer to the NDS Chap. 6.

PROBLEM 47

A normalweight (0.150 kip/ft^3) reinforced concrete retaining wall is designed to support the loads shown. An 8 ft wide surcharge extends along the entire length of the retaining wall. The soil is sandy without silt. The soil behind the retaining wall is soil 1. The soil in front of the retaining wall is soil 2.

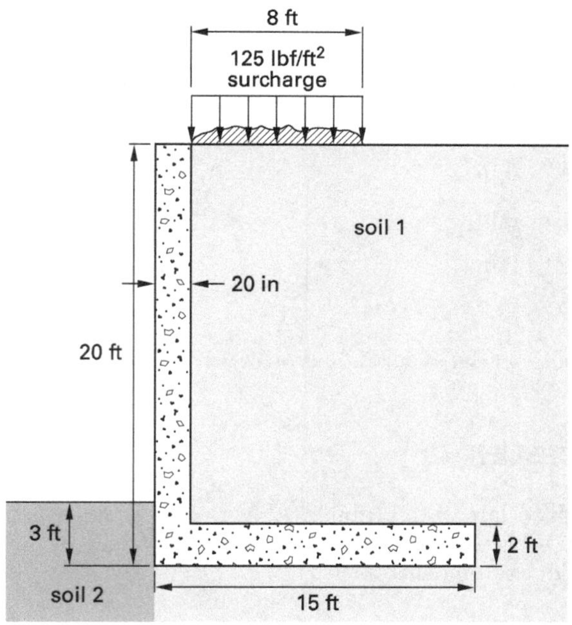

characteristic	soil 1	soil 2
unit weight (kip/ft^3)	0.110	0.100
angle of internal friction (degrees)	28	15
cohesion (kip/ft^2)	0	0.300
active earth pressure coefficient	0.361	0.589
passive earth pressure coefficient	2.77	1.70

Design & Details of Structures

The factor of safety against sliding is

(A) 1.03

(B) 1.11

(C) 1.44

(D) 1.66

Hint: Passive restraint from the soil is only considered if it will always be there. In most cases, it is neglected.

MATERIALS AND MATERIAL PROPERTIES

PROBLEM 48

A structural masonry wall is designed to span 10 ft vertically from a structure's foundation to the roof. The wall is not reinforced but is grouted solid and carries the load from the roof. The compressive strength of masonry is not specified. Lateral support is provided by intersecting walls spaced 15 ft on center. Using empirical design, what is most nearly the minimum thickness of the wall?

(A) 6 in

(B) 8 in

(C) 9 in

(D) no limit

Hint: A structural masonry wall that supports an axial load (i.e., roof load) is a bearing wall.

PROBLEM 49

A glued laminated (glulam) beam with the designation 20F-V7 is used to span a 20 ft opening. Which of the following statements is true?

(A) The beam is mechanically graded.

(B) The depth of the beam is 20 in.

(C) The shear capacity of the beam is 700 lbf/in^2.

(D) The flexural tensile capacity of the beam is 2000 lbf/in^2.

Hint: The answer does not depend upon the type of wood.

PROBLEM 50

A built-up column made from three southern pine sawn 2 × 6 members nailed together meets the requirements of NDS Sec. 15.3.3. The ends of the column are free to rotate but are not free to translate.

distance between points of lateral support of compression member in plane 1	9.0 ft
distance between points of lateral support of compression member in plane 2	4.5 ft
E'_{min}	620,000 lbf/in^2
F_c^*	1750 lbf/in^2

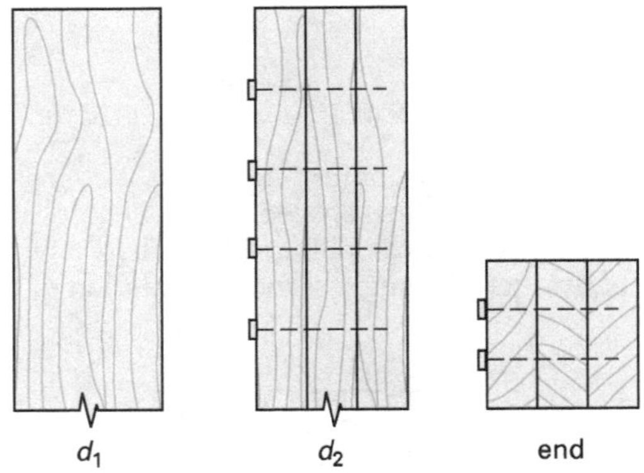

What is the column stability factor in direction d_1 for this column?

(A) 0.38

(B) 0.52

(C) 0.59

(D) 0.65

Hint: NDS Sec. 15.3 covers the design of built-up columns.

PROBLEM 51

An elementary school is constructed with hollow brick bearing walls with a specified compressive strength of masonry of 4950 lbf/in^2. The brick meets the requirements of ASTM C652, and Type S mortar is used. What is most nearly the modulus of elasticity of the masonry?

(A) 1.4×10^6 lbf/in^2

(B) 2.7×10^6 lbf/in^2

(C) 3.5×10^6 lbf/in^2

(D) 4.1×10^6 lbf/in^2

Hint: ASTM C652 is for clay brick.

PROBLEM 52

Which of the following statements are true?

I. Mat foundations can be used in areas where the basement is below the ground water table (GWT).

II. Mat foundations always require a top layer of reinforcing bars.

III. Conventional spread footings tolerate larger differential settlements than do mat foundations.

IV. Mat foundations are not suitable where settlement may be a problem.

(A) I only

(B) I and II

(C) III and IV

(D) II, III, and IV

Hint: A mat foundation is a large concrete slab in contact with the soil and is commonly used to support several columns or pieces of equipment.

PROBLEM 53

$^3/_4$ in diameter A307 headed anchor bolts are used to attach a steel angle to a grouted concrete masonry wall. The anchor bolts have an effective embedment length of 6 in, and edge distance is not a concern. The specified compressive strength of masonry is 2000 lbf/in^2. Using allowable stress design, the allowable shear load on the bolts is most nearly

(A) 1700 lbf

(B) 1900 lbf

(C) 5700 lbf

(D) 12,600 lbf

Hint: See TMS 402 Sec. 8.1.3.

PROBLEM 54

A concrete slab designed for a parking garage constructed in Chicago uses normalweight concrete with a compressive strength of 3000 lbf/in^2. The maximum aggregate size is 1 in. The total percentage of air content of the concrete mix is most nearly

(A) 0.06%

(B) 0.45%

(C) 4.5%

(D) 6.0%

Hint: The parking garage in Chicago is exposed to freezing and thawing conditions. Refer to ACI Sec. 19.3.

PROBLEM 55

Using the *National Design Specification for Wood Construction* (NDS), which of the following statements is true?

I. The temperature factor, C_t, applies to members that are subjected to extremely cold temperatures.

II. The volume factor, C_V, applies only to glued laminated timber and structural composite lumber bending members.

III. The bending design allowable stress, F_b, for a floor framed with 1 × 6 sawn lumber joists must be multiplied by the repetitive member factor.

IV. The load duration factor, C_D, does not apply to the modulus of elasticity values.

(A) I and II

(B) II and III

(C) II and IV

(D) III and IV

Hint: Refer to NDS Sec. 2.3 for adjustment of design values.

Design & Details of Structures

PROBLEM 56

The W18 × 40 composite steel beam shown spans 30 ft and supports a 4 in concrete slab. The modular ratio, n, is 8.

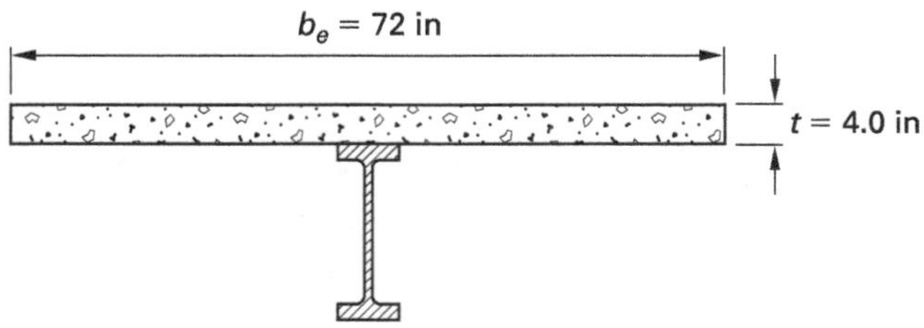

If the concrete slab is transformed to steel, the distance to the centroid of the transformed section measured from the bottom of the beam will be most nearly

(A) 14 in

(B) 16 in

(C) 17 in

(D) 19 in

Hint: The modular ratio, n, is the ratio E_s/E_c.

PROBLEM 57

A prestressed concrete beam has a specified compressive strength of concrete of 6000 lbf/in^2 and the following properties.

compressive strength of concrete at time of initial prestress	4000 lbf/in^2
specified yield strength of nonprestressed reinforcement	60,000 lbf/in^2
initial prestress force	150 kips

The extreme fiber stress in tension in the concrete immediately after prestress transfer is limited to

(A) 150 lbf/in^2

(B) 190 lbf/in^2

(C) 230 lbf/in^2

(D) 380 lbf/in^2

Hint: Refer to ACI 318 Chap. 24.5.

SOLUTION 1

ACI 318 Sec. 25.5.2 contains the lap splice requirements for deformed bars in tension. ACI Table 25.5.2.1 gives the required length of a lap splice as a function of the splice type, which is either class A or class B.

Splice types are described in ACI Table 10.7.5.2.2. If the provided area of reinforcement is at least twice the area of reinforcement required by analysis at the splice location, and no more than 50% of the steel is spliced within the required lap length, a class A may be used; otherwise, a class B must be used.

In this case, four no. 7 bars are provided at the splice location, only two no. 7 bars are required, and only half the bars are spliced, so a class A lap splice may be used. From ACI Table 25.5.2.1, the length of a class A lap splice is the greater of 12 in or $1.0l_d$.

The required development length, l_d, of reinforcing bars in tension is given in ACI Sec. 25.4.2 and is a function of the spacing and cover of the reinforcement. Number 7 bars are used, and they have a diameter of 0.875 in. The clear spacing of the reinforcement, given as 4 in, is greater than twice the diameter of the reinforcement, and the clear cover is greater than the diameter of the bars. For no. 7 bars meeting these criteria for spacing and cover, ACI Table 25.4.2.2 gives the development length as

$$l_d = \left(\frac{f_y\psi_t\psi_c}{20\lambda\sqrt{f'_c}}\right)d_b$$

From ACI Sec. 25.4.2.1, the development length must also be at least 12 in.

From ACI Table 25.4.2.4, the modification factor λ is 1.0 for normalweight concrete, and the modification factor ψ_c is 1.0 for uncoated bars. The bars are located at the top of the beam, and the depth of the beam is 20 in, so more than 12 in of fresh concrete is placed below the reinforcement. From ACI Table 25.4.2.4, then, the modification factor ψ_t is 1.3. The development length is

$$\begin{aligned} l_d &= \left(\frac{f_y\psi_t\psi_c}{20\lambda\sqrt{f'_c}}\right)d_b \geq 12 \text{ in} \\ &= \left(\frac{\left(60{,}000\ \frac{\text{lbf}}{\text{in}^2}\right)(1.3)(1.0)}{(20)(1.0)\sqrt{3000\ \frac{\text{lbf}}{\text{in}^2}}}\right)(0.875 \text{ in}) \\ &= 62.3 \text{ in} \end{aligned}$$

The minimum required lap splice length is

$$\begin{aligned} l_{st} &= 1.0 l_d \geq 12 \text{ in} \\ &= (1.0)(62.3 \text{ in}) \\ &= 62.3 \text{ in} \end{aligned}$$

This is a minimum, so round up to 63 in.

The answer is (B).

Why Other Options Are Wrong

(A) This incorrect solution uses a yield strength of reinforcement of 40,000 lbf/in^2 instead of 60,000 lbf/in^2.

(C) This incorrect solution is the lap splice length for a class B splice, not class A.

(D) This incorrect solution uses the equation from ACI Table 25.4.2.2 for the development length for "other cases."

SOLUTION 2

Use the column base plates design procedure given in Sec. J.8 of the AISC *Steel Construction Manual.*

For a W12 × 72 steel column,

$$\begin{aligned} d &= 12.25 \text{ in} \\ b_f &= 12.0 \text{ in} \end{aligned}$$

Determine the available bearing strength, where ϕ_c is 0.65.

$$\phi_c P_p = \phi_c 0.85 f_c' A_1 \sqrt{\frac{A_2}{A_1}}$$

$$A_1 = NB = (16 \text{ in})(14 \text{ in}) = 224 \text{ in}^2$$

$$\begin{aligned} A_2 = L_{\text{ftg}} W_{\text{ftg}} &= \left[(8 \text{ ft})\left(12 \ \frac{\text{in}}{\text{ft}}\right)\right]\left[(8 \text{ ft})\left(12 \ \frac{\text{in}}{\text{ft}}\right)\right] \\ &= 9216 \text{ in}^2 \end{aligned}$$

$$\sqrt{\frac{A_2}{A_1}} = \sqrt{\frac{9216 \text{ in}^2}{224 \text{ in}^2}} = 6.41 \quad [> 2.0. \text{ Use } 2.0.]$$

$$\begin{aligned} \phi_c P_p &= \phi_c 0.85 f_c' A_1 \sqrt{\frac{A_2}{A_1}} \\ &= (0.65)(0.85)\left(3.0 \ \frac{\text{kips}}{\text{in}^2}\right)(224 \text{ in}^2)(2.0) \\ &= 743 \text{ kips} \end{aligned}$$

$$P_u = 630 \text{ kips} < \phi_c P_p = 743 \text{ kips} \quad [\text{OK}]$$

The thickness of the base plate is

$$t_p = l \sqrt{\frac{2P_u}{0.9 F_y BN}}$$

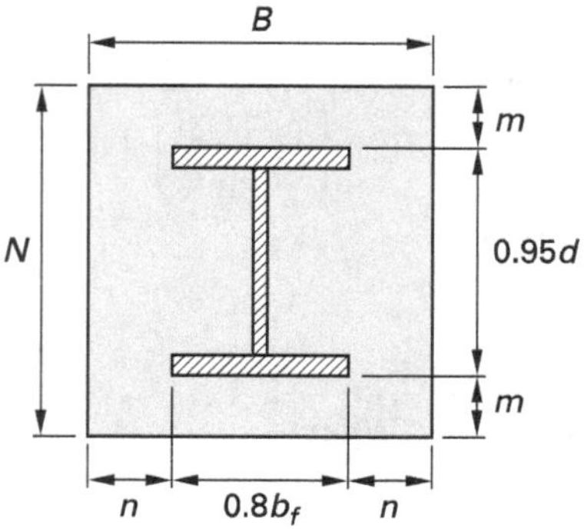

Determine the critical cantilever projection of the base plate, l. Use the larger of m, n, and $\lambda n'$.

$$m = \frac{N - 0.95d}{2} = \frac{16 \text{ in} - (0.95)(12.25 \text{ in})}{2} = 2.18 \text{ in}$$

$$n = \frac{B - 0.80 b_f}{2} = \frac{14 \text{ in} - (0.80)(12.0 \text{ in})}{2} = 2.20 \text{ in}$$

$$n' = \frac{\sqrt{d b_f}}{4} = \frac{\sqrt{(12.25 \text{ in})(12.0 \text{ in})}}{4} = 3.03 \text{ in}$$

Determine λ.

$$\lambda = \frac{2\sqrt{X}}{1 + \sqrt{1 - X}} \leq 1.0$$

$$\begin{aligned} X &= \frac{4 d b_f P_u}{(d + b_f)^2 (\phi_c P_p)} \\ &= \frac{(4)(12.25 \text{ in})(12.0 \text{ in})(630 \text{ kips})}{(12.25 \text{ in} + 12.0 \text{ in})^2 (743 \text{ kips})} \\ &= 0.848 \end{aligned}$$

$$\begin{aligned} \lambda &= \frac{2\sqrt{X}}{1 + \sqrt{1 - X}} \\ &= \frac{2\sqrt{0.848}}{1 + \sqrt{1 - 0.848}} \\ &= 1.33 \quad [> 1.0. \text{ Use } 1.0.] \\ \lambda n' &= (1.0)(3.03 \text{ in}) \\ &= 3.03 \text{ in} \quad [\text{governs}] \end{aligned}$$

Determine the required base plate thickness, where $l = \lambda n'$.

$$t_{req} = l\sqrt{\frac{2P_u}{0.9F_yBN}} = 3.03\ \text{in}\sqrt{\frac{(2)(630\ \text{kips})}{(0.9)\left(36\ \frac{\text{kips}}{\text{in}^2}\right)(14\ \text{in})(16\ \text{in})}} = 1.26\ \text{in}$$

The answer is (C).

Why Other Options Are Wrong

(A) This incorrect solution neglects to calculate $\lambda n'$, which in this case is the controlling distance.

(B) This incorrect solution uses F_u instead of F_y when calculating the required thickness.

(D) This incorrect solution reverses N and B when calculating m and n. N is the length of the base plate in the direction of the column depth. B is the width of the base plate in the direction of the column flange, as shown in Fig. 14-3 of the AISC column base plates design procedure.

SOLUTION 3

Determine the design loads.

$$\begin{aligned} P_u &= 1.2D + 1.6L \\ &= (1.2)(130\ \text{kips}) + (1.6)(390\ \text{kips}) \\ &= 780\ \text{kips} \\ P_u &\leq \phi_c P_n \\ 780\ \text{kips} &\leq \phi_c P_n \end{aligned}$$

The values in the columns tables in the AISC *Steel Construction Manual* are based on an effective length with respect to the minor axis, KL_y.

The effective column lengths are given as

$$\begin{aligned} KL_x &= 32\ \text{ft} \\ KL_y &= 18\ \text{ft} \end{aligned}$$

Use AISC Table 4-1. Since the column depth is limited to 12 in, begin with the W12 sections. The effective length is 18 ft. For $\phi_c P_n$ (LRFD), use the unshaded columns and determine that a W12 × 87 column has a capacity of 802 kips. Check the capacity based on the effective length for buckling about the x-axis.

From the AISC column tables, $r_x/r_y = 1.75$.

The equivalent effective length for the x-axis is

$$\frac{L_x}{\frac{r_x}{r_y}} = \frac{32\ \text{ft}}{1.75} = 18.3\ \text{ft} \quad [> 18\ \text{ft}]$$

Therefore, check the capacity for $KL_y = 18.3$ ft.

By interpolation, a W12 × 87 column has a capacity of 792 kips [OK].

The answer is (A).

Why Other Options Are Wrong

(B) This solution incorrectly uses the column for ASD instead of LRFD in the column tables to find that a W12 × 136 column is needed.

(C) This incorrect solution reverses the major and minor axes in determining effective length.

(D) This incorrect solution reverses the major and minor axes in determining effective length and uses the column for ASD instead of LRFD in the column tables.

SOLUTION 4

The area of the square footing is

$$A = BL = (6.5\ \text{ft})(6.5\ \text{ft}) = 42.25\ \text{ft}^2$$

The ultimate load on the footing is the larger of

$$\begin{aligned} P_u &= 1.4P_D = (1.4)(50\ \text{kips}) = 70\ \text{kips} \\ P_u &= 1.2P_D + 1.6P_L \\ &= (1.2)(50\ \text{kips}) + (1.6)(75\ \text{kips}) \\ &= 180\ \text{kips} \quad [\text{controls}] \end{aligned}$$

The factored soil pressure is

$$q_u = \frac{P_u}{A} = \frac{180\ \text{kips}}{42.25\ \text{ft}^2} = 4.26\ \text{kips/ft}^2$$

The ultimate two-way punching shear is

$$\begin{aligned} V_u = P_u - q_u A_{\text{critical}} &= 180\ \text{kips} - \frac{\left(4.26\ \frac{\text{kips}}{\text{ft}^2}\right)(150\ \text{in}^2)}{144\ \frac{\text{in}^2}{\text{ft}^2}} \\ &= 175.6\ \text{kips} \quad (180\ \text{kips}) \end{aligned}$$

The answer is (D).

Why Other Options Are Wrong

(A) This incorrect solution calculates the punching shear based on the factored soil pressure on the critical area only.

(B) This incorrect solution does not use factored loads when calculating the load on the footing.

(C) This incorrect solution makes an error in the conversion of the critical area.

SOLUTION 5

The size of the footing is based on service (unfactored) loads and soil pressures, because footing design safety is provided by the safety factor in the allowable soil bearing pressure.

$$P = P_D + P_L$$
$$M = M_D + M_L$$

For column 1,

$$P_1 = 30 \text{ kips} + 60 \text{ kips} = 90 \text{ kips}$$
$$M_1 = 30 \text{ ft-kips} + 40 \text{ ft-kips} = 70 \text{ ft-kips}$$

For column 2,

$$P_2 = 60 \text{ kips} + 60 \text{ kips} = 120 \text{ kips}$$
$$M_2 = 30 \text{ ft-kips} + 40 \text{ ft-kips} = 70 \text{ ft-kips}$$

For the soil pressure under the footing to be uniform, the resultant load, R, must be located at the centroid of the base area.

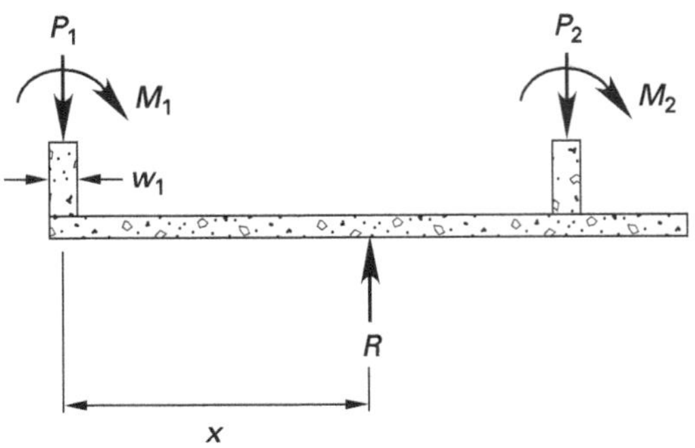

Summing moments about column 1, x is the distance to the centroid of the footing.

$$\begin{aligned} Rx &= P_2(18 \text{ ft}) + M_1 + M_2 \\ (P_1 + P_2)x &= P_2(18 \text{ ft}) + M_1 + M_2 \\ (90 \text{ kips} + 120 \text{ kips})x &= (120 \text{ kips})(18 \text{ ft}) + 70 \text{ ft-kips} \\ &\quad + 70 \text{ ft-kips} \\ (210 \text{ kips})x &= 2300 \text{ ft-kips} \\ x &= 11.0 \text{ ft} \end{aligned}$$

The length of the footing is

$$\begin{aligned} L &= \left(\tfrac{1}{2}w_1 + x\right)(2) = \left(\left(\frac{1}{2}\right)(1 \text{ ft}) + 11.0 \text{ ft}\right)(2) \\ &= 23.0 \text{ ft} \end{aligned}$$

The answer is (D).

Why Other Options Are Wrong

(A) This incorrect solution neglects the effects of the applied moments in calculating the location of the resultant load.

(B) This incorrect solution neglects to add one-half the column width when calculating the length of the footing.

(C) This incorrect solution uses factored loads instead of service loads to size the footing. Factored loads are used in the design of the reinforcement for the concrete footing. The size of the footing is based on service (unfactored) loads and soil pressures, because footing design safety is provided by the safety factor in the allowable soil bearing pressure.

SOLUTION 6

The load from the joist is assumed to act at the point of the anchorage. If the load from the joists acts within the center third of the wall cross section, the eccentricity from the load is negligible. For a 12 in wall, the specified dimension is 11.63 in. If the eccentricity does not exceed 1.9 in, the load acts within the center third.

$$e = \frac{11.63 \text{ in}}{2} - 2 \text{ in} = 3.82 \text{ in}$$

The eccentricity of the load must be considered.

If the load acts at 12 in on center, calculate the properties of the wall.

$$A = bh = (12 \text{ in})(11.63 \text{ in}) = 140 \text{ in}^2$$
$$S = \frac{bh^2}{6} = \frac{(12 \text{ in})(11.63 \text{ in})^2}{6} = 271 \text{ in}^3$$

The axial stress on the wall is

$$f_a = \frac{P}{A} = \frac{700 \text{ lbf}}{140 \text{ in}^2} = 5.0 \text{ lbf/in}^2$$

The bending stress on the wall creates tension on the outside face and compression on the inside face of the wall.

$$f_b = \frac{Pe}{S} = \frac{(700 \text{ lbf})(3.82 \text{ in})}{271 \text{ in}^3} = 9.87 \text{ lbf/in}^2$$

Design & Details of Structures

The total stress on the wall is

$$f_a + f_b = 5.0\ \frac{\text{lbf}}{\text{in}^2} + 9.87\ \frac{\text{lbf}}{\text{in}^2} = 14.9\ \text{lbf/in}^2 \quad (15\ \text{lbf/in}^2\ \text{(compression)})$$

$$f_a + f_b = 5.0\ \frac{\text{lbf}}{\text{in}^2} - 9.87\ \frac{\text{lbf}}{\text{in}^2} = -4.87\ \text{lbf/in}^2 \quad (5.0\ \text{lbf/in}^2\ \text{(tension)})$$

The answer is (D).

Why Other Options Are Wrong

(A) This incorrect solution neglects the effects of the eccentricity of the load (i.e., it does not calculate the bending stresses).

(B) This incorrect solution mistakenly identifies the axial stress on the wall as a tensile stress and does not consider the flexural stresses on the wall.

(C) This incorrect solution calculates the bending stresses due to the eccentricity of the load but does not add the flexural compressive stress to the axial compressive stress.

SOLUTION 7

Section 24.2.3 of ACI 318 contains the requirements for deflection calculations for beams. Immediate deflection calculations are based on service (unfactored) loads and the effective moment of inertia, I_e.

$$\Delta = \frac{5wL^4}{384E_cI_e}$$

The total uniform load is

$$\begin{aligned} w &= D + L + \text{beam self-weight} \\ &= 800\ \frac{\text{lbf}}{\text{ft}} + 1200\ \frac{\text{lbf}}{\text{ft}} \\ &\quad + (16\ \text{in})(30\ \text{in})\left(\frac{1\ \text{ft}}{12\ \text{in}}\right)^2\left(145\ \frac{\text{lbf}}{\text{ft}^3}\right) \\ &= 2483\ \text{lbf/ft} \end{aligned}$$

$$\begin{aligned} E_c &= 57{,}000\sqrt{f_c'} = 57{,}000\sqrt{6000\ \frac{\text{lbf}}{\text{in}^2}} \\ &= 4.42 \times 10^6\ \text{lbf/in}^2 \quad \text{[ACI 318 Sec. 19.2.2.1.b]} \end{aligned}$$

$$I_e = \left(\frac{M_{cr}}{M_a}\right)^3 I_g + \left(1 - \left(\frac{M_{cr}}{M_a}\right)^3\right) I_{cr} \quad \text{[ACI 318 Eq. 24.2.3.5a]}$$

$$I_g = \frac{bh^3}{12} = \frac{(16\ \text{in})(30\ \text{in})^3}{12} = 36{,}000\ \text{in}^4$$

$$M_{cr} = 110\ \text{ft-kips}$$

$$\begin{aligned} M_a &= \frac{wL^2}{8} = \frac{\left(2483\ \frac{\text{lbf}}{\text{ft}}\right)(24\ \text{ft})^2\left(1000\ \frac{\text{lbf}}{\text{kip}}\right)}{8} \\ &= 179\ \text{ft-kips} \end{aligned}$$

$$I_{cr} = 15{,}500\ \text{in}^4$$

$$\begin{aligned} I_e &= \left(\frac{M_{cr}}{M_a}\right)^3 I_g + \left(1 - \left(\frac{M_{cr}}{M_a}\right)^3\right) I_{cr} \\ &= \left(\frac{110\ \text{ft-kips}}{179\ \text{ft-kips}}\right)^3 (36{,}000\ \text{in}^4) \\ &\quad + \left(1 - \left(\frac{110\ \text{ft-kips}}{179\ \text{ft-kips}}\right)^3\right)(15{,}500\ \text{in}^4) \\ &= 20{,}260\ \text{in}^4 \end{aligned}$$

The immediate deflection is

$$\begin{aligned} \Delta &= \frac{5wL^4}{384E_cI_e} = \frac{(5)\left(2483\ \frac{\text{lbf}}{\text{ft}}\right)(24\ \text{ft})^4\left(12\ \frac{\text{in}}{\text{ft}}\right)^3}{(384)\left(4.42 \times 10^6\ \frac{\text{lbf}}{\text{in}^2}\right)(20{,}260\ \text{in}^4)} \\ &= 0.21\ \text{in} \end{aligned}$$

The answer is (C).

Why Other Options Are Wrong

(A) This incorrect solution uses the gross moment of inertia to calculate the deflection. This does not account for the effects of cracking.

(B) This incorrect solution does not include the beam self-weight in the calculation of the loads.

(D) This incorrect solution uses factored loads in the calculations and does not include the beam self-weight. Immediate deflection calculations should be based on unfactored loads.

SOLUTION 8

Section 4 of the AISC *Steel Construction Manual* covers the design of compression members. The available strength in axial compression, AISC *Manual* Table 4-8, can be used to find the critical compressive stress for the double angle in this case.

For an L6 × 6 × ½ double angle with Kl_x of 24 ft,

$$\phi_c P_n = 108 \text{ kips}$$

The critical compressive stress is

$$F_{cr} = \frac{P_n}{A_g} = \frac{108 \text{ kips}}{(0.9)(11.5 \text{ in}^2)} = 10.4 \text{ kips/in}^2 \quad (10 \text{ kips/in}^2)$$

Alternate Solution

The critical compressive stress can also be determined using the AISC *Specification.* To prevent localized buckling, AISC *Specification* Sec. B4 limits the width-to-thickness ratio so that the plate element is fully effective. For the double-angle member shown, AISC *Specification* Table B4.1 gives

$$\frac{b}{t} \leq 0.45\sqrt{\frac{E}{F_y}} = 0.45\sqrt{\frac{29{,}000 \frac{\text{kips}}{\text{in}^2}}{36 \frac{\text{kips}}{\text{in}^2}}} = 12.77$$

$$\frac{b}{t} = \frac{6 \text{ in}}{0.5 \text{ in}} = 12.0$$

Since 12.0 < 12.77, the section is not classified as a slender element and can be designed in accordance with AISC *Specification* Sec. B and Sec. E3.

AISC *Specification* Sec. E3 gives the requirements for unstiffened compression elements without slender elements such as the angles shown.

The allowable compressive stress depends on Kl/r relative to

$$4.71\sqrt{\frac{E}{F_y}} = 4.71\sqrt{\frac{29{,}000 \frac{\text{kips}}{\text{in}^2}}{36 \frac{\text{kips}}{\text{in}^2}}} = 134$$

$$\frac{Kl}{r} = 155 \quad [> 134]$$

Use AISC *Specification* Eq. E3-3 to find the allowable compressive stress.

$$F_{cr} = 0.877F_e$$

The elastic critical buckling stress, F_e, is determined by AISC *Specification* Eq. E3-4.

$$F_e = \frac{\pi^2 E}{\left(\frac{Kl}{r}\right)^2} = \frac{\pi^2\left(29{,}000 \frac{\text{kips}}{\text{in}^2}\right)}{(155)^2} = 11.9 \text{ kips/in}^2$$

The allowable compressive stress is

$$F_{cr} = 0.877F_e = (0.877)\left(11.9 \frac{\text{kips}}{\text{in}^2}\right) = 10.4 \text{ kips/in}^2 \quad (10 \text{ kips/in}^2)$$

The answer is (C).

Why Other Options Are Wrong

(A) This incorrect solution solves the problem correctly but identifies the wrong units for the critical stress.

(B) This incorrect solution finds the stress by dividing the tabulated available strength value by the gross area, neglecting the safety factor.

(D) This incorrect solution finds the criticial stress for a member made of ASTM A992 steel instead of ASTM A36 steel.

SOLUTION 9

The design of bearing stiffeners is found in AASHTO Sec. 6.10.11.2. According to this section, the factored axial resistance, P_r, is determined in accordance with AASHTO Sec. 6.9.2.1. Since the bearing stiffener is a non-composite (steel only) element, use Sec. 6.9.4.

$$P_r = \phi_c P_n \quad \text{[AASHTO Eq. 6.9.2.1-1]}$$

From AASHTO Sec. 6.5.4.2, the resistance factor for axial compression, ϕ_c, is 0.9. The selection of the equation for nominal axial resistance, P_n, depends on the ratio of P_e/P_o. Using AASHTO Table 6.9.4.1.1-1, the elastic critical buckling resistance, P_e, is found as

$$P_e = \left(\frac{\pi^2 E}{\left(\frac{Kl}{r_s}\right)^2}\right)A_g \quad \text{[AASHTO Eq. 6.9.4.1.2-1]}$$

$$P_o = QF_yA_g$$

According to Sec. C6.9.4.1.1, Q is always taken as 1.0 for bearing stiffeners.

Determine the properties of the bearing stiffeners.

Check that the minimum width of the stiffener, b_t, of 6 in meets the AASHTO requirement. The thickness of the projecting stiffener element, t_p, is given as 0.75 in.

$$b_t \le 0.48t_p\sqrt{\frac{E}{F_{ys}}} \quad \text{[AASHTO Eq. 6.10.11.2.2-1]}$$

$$= (0.48)(0.75\ \text{in})\sqrt{\frac{29{,}000\ \frac{\text{kips}}{\text{in}^2}}{50\ \frac{\text{kips}}{\text{in}^2}}}$$

$$= 8.67\ \text{in} \quad [b_t = 6\ \text{in, OK}]$$

The effective web width is given in AASHTO Sec. 6.10.11.2.4b as

$$w = 2(9t_w) = (2)\big((9)(0.5\ \text{in})\big) = 9\ \text{in}$$

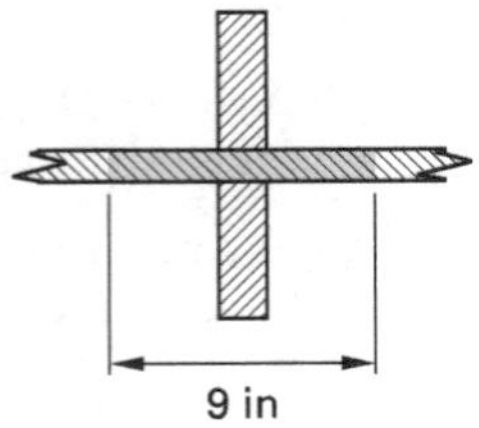

The cross-sectional area of the bearing stiffener column section is the sum of the area of the bearing stiffener and the area of the beam web.

$$A_s = 2b_tt_p + wt_{\text{web}}$$

$$= (2)(6\ \text{in})(0.75\ \text{in}) + (9\ \text{in})(0.5\ \text{in})$$

$$= 13.5\ \text{in}^2$$

Determine the moment of inertia about the web centerline.

$$I = \frac{bh^3}{12} = \frac{t_p(b_t + t_{\text{web}} + b_t)^3}{12} + \frac{(w - t_p)(t_{\text{web}})^3}{12}$$

$$= \frac{(0.75\ \text{in})(6\ \text{in} + 0.5\ \text{in} + 6\ \text{in})^3}{12} + \frac{(9\ \text{in} - 0.75\ \text{in})(0.5\ \text{in})^3}{12}$$

$$= 122\ \text{in}^4$$

The radius of gyration for the steel section is

$$r_s = \sqrt{\frac{I}{A_s}} = \sqrt{\frac{122\ \text{in}^4}{13.5\ \text{in}^2}} = 3.01\ \text{in}$$

Determine the column's slenderness ratio. From AASHTO Sec. 6.10.11.2.4a, the effective length factor, K, equals 0.75.

$$\frac{Kl}{r_s} = \frac{(0.75)(7.0\ \text{ft})\left(12\ \frac{\text{in}}{\text{ft}}\right)}{3.01\ \text{in}} = 20.9$$

The elastic critical buckling resistance is

$$P_e = \left(\frac{\pi^2 E}{\left(\frac{Kl}{r_s}\right)^2}\right)A_g$$

$$= \frac{\pi^2\left(29{,}000\ \frac{\text{kips}}{\text{in}^2}\right)(13.5\ \text{in}^2)}{(20.9)^2}$$

$$= 8845\ \text{kips}$$

The equivalent nominal yield resistance is

$$P_o = QF_yA_g = (1.0)\left(50\ \frac{\text{kips}}{\text{in}^2}\right)(13.5\ \text{in}^2)$$

$$= 675\ \text{kips}$$

Check the nominal compressive resistance criteria per AASHTO Sec. 6.9.4.1.1.

$$\frac{P_e}{P_o} = \frac{8845\ \text{kips}}{675\ \text{kips}} = 13.1 > 0.44$$

Use AASHTO Eq. 6.9.4.1.1-1.

$$\frac{P_o}{P_e} = \frac{675\ \text{kips}}{8845\ \text{kips}} = 0.076$$

The nominal compressive resistance is

$$P_n = (0.658^{P_o/P_e})P_o = (0.658^{0.076})(675\ \text{kips})$$

$$= 653.9\ \text{kips}$$

For the column section, the factored axial resistance of the effective bearing stiffener is

$$P_r = \phi_c P_n = (0.9)(653.9\ \text{kips})$$

$$= 588\ \text{kips} \quad (590\ \text{kips})$$

The answer is (B).

Why Other Options Are Wrong

(A) This incorrect solution uses a value of 1.0 for K instead of 0.75.

(C) This incorrect solution finds the nominal resistance instead of the factored resistance.

(D) This incorrect solution mistakenly calculates the factored axial resistance as P_n/ϕ.

SOLUTION 10

Section I of the AISC *Steel Construction Manual* governs composite design. Section I3.1b indicates that for unshored construction, the steel alone must carry the loads prior to concrete hardening. According to Sec. F2, the maximum strength is obtained when $L_b \leq L_p$.

$$M_u = \phi_b M_p$$

Use Table 3-19, Composite W Shapes Available Strength in Flexure. For a W21 × 55 composite beam,

$$\phi_b M_p = 473 \text{ ft-kips} \quad (470 \text{ ft-kips})$$

The answer is (C).

Why Other Options Are Wrong

(A) This incorrect solution uses the column for ASD instead of LRFD in the Available Strength table.

(B) This incorrect solution multiplies the ASD value in the Available Strength table by $\phi_b = 0.90$.

(D) This incorrect solution uses the value for a W21 × 57 instead of a W21 × 55 in the Available Strength table.

SOLUTION 11

Per ACI 318 Sec. 22.5.5.1, the nominal shear strength provided by the concrete alone is

$$\begin{aligned} V_c &= 2\sqrt{f'_c}\, b_w d = 2\sqrt{3000 \frac{\text{lbf}}{\text{in}^2}}(6 \text{ in})(7.5 \text{ in}) \\ &= 4929.5 \text{ lbf} \\ \phi V_c &= 0.75 V_c = (0.75)(4929.5 \text{ lbf}) \\ &= 3697 \text{ lbf} \end{aligned}$$

Based on ACI 318 Sec. 9.6.3.1, a beam with depth of less than 10 in and $0.5\phi V_c < V_u < \phi V_c$ requires no minimum shear reinforcement.

$$\begin{aligned} 0.5\phi V_c &= (0.5)(3697 \text{ lbf}) = 1849 \text{ lbf} \\ V_u &= 2000 \text{ lbf} \quad [> 1849 \text{ lbf and } < 3697 \text{ lbf}] \\ h &= 9 \text{ in} \quad [< 10 \text{ in}] \end{aligned}$$

Since both conditions are met, no shear reinforcement is required.

The answer is (D).

Why Other Options Are Wrong

(A) This incorrect solution misses the exception, given in ACI 318 Sec. 9.6.3.1, to the minimum shear requirement and mistakenly uses the minimum shear reinforcement requirement found in Sec. 9.7.6.2.2.

(B) This incorrect solution misses the exception, given in ACI 318 Sec. 9.6.3.1, to the minimum shear requirement and uses h for d when calculating the maximum stirrup spacing.

(C) This incorrect solution misses the exception, given in ACI 318 Sec. 9.6.3.1, to the minimum shear requirement and incorrectly uses the largest stirrup spacing found in ACI 318 Sec. 9.7.6.2.2 instead of the smallest.

SOLUTION 12

Determine the effective depth to reinforcement, d. The two bars side-by-side at the bottom of the beam are shown.

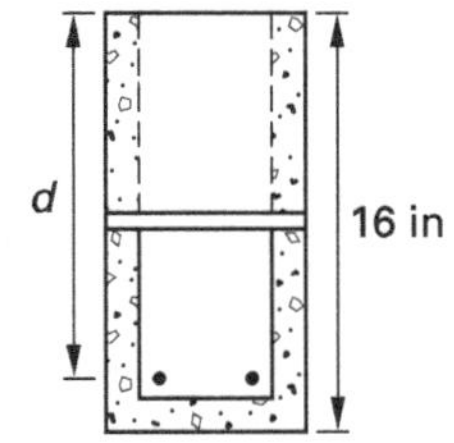

Allowing for the thickness of the unit (1.25 in) and one-half the bar diameter (0.25 in) plus grout cover (0.5 in), estimate d.

$$d = 16 \text{ in} - (1.25 \text{ in} + 0.25 \text{ in} + 0.5 \text{ in}) = 14 \text{ in}$$

Determine the stress in the steel using the equations found in a reference handbook on masonry walls.

$$f_s = \frac{M}{A_s jd}$$

$$\rho = \frac{A_s}{bd}$$

$$\begin{aligned} k &= \sqrt{2\rho n + (\rho n)^2} - \rho n \\ j &= 1 - \frac{k}{3} \\ n &= \frac{E_s}{E_m} \end{aligned}$$

Building Code Requirements for Masonry Structures (TMS 402) Sec. 4.2.2 specifies the moduli of elasticity of steel and masonry, respectively, as

$$E_s = 29 \times 10^6 \text{ lbf/in}^2$$

$$\begin{aligned} E_m &= 900f'_m \quad \text{[for concrete masonry]} \\ &= (900)\left(3000 \ \frac{\text{lbf}}{\text{in}^2}\right) \\ &= 2.7 \times 10^6 \text{ lbf/in}^2 \end{aligned}$$

$$\begin{aligned} n &= \frac{E_s}{E_m} = \frac{29 \times 10^6 \ \frac{\text{lbf}}{\text{in}^2}}{2.7 \times 10^6 \ \frac{\text{lbf}}{\text{in}^2}} \\ &= 10.7 \end{aligned}$$

Given two no. 4 bars, App. E of ACI 318 lists the area of a no. 4 reinforcing bar as 0.20 in^2.

$$A_{s,\text{prov}} = (2)(0.20 \text{ in}^2) = 0.40 \text{ in}^2$$

$$\begin{aligned} \rho &= \frac{A_{s,\text{prov}}}{bd} = \frac{0.40 \text{ in}^2}{(7.5 \text{ in})(14 \text{ in})} \\ &= 0.00381 \end{aligned}$$

$$\begin{aligned} k &= \sqrt{2\rho n + (\rho n)^2} - \rho n \\ &= \sqrt{(2)(0.00381)(10.7) + ((0.00381)(10.7))^2} \\ &\quad -(0.00381)(10.7) \\ &= 0.248 \end{aligned}$$

$$\begin{aligned} j &= 1 - \frac{k}{3} = 1 - \frac{0.248}{3} \\ &= 0.917 \end{aligned}$$

The stress in the steel is

$$\begin{aligned} f_s &= \frac{M}{A_s jd} = \frac{88{,}000 \text{ in-lbf}}{(0.40 \text{ in}^2)(0.917)(14 \text{ in})} \\ &= 17{,}137 \text{ lbf/in}^2 \quad (17{,}000 \text{ lbf/in}^2) \end{aligned}$$

The allowable tensile stress in the steel is given in TMS 402/602 Sec. 8.3.3 for grade 60 reinforcement.

$$F_s = 32{,}000 \text{ lbf/in}^2 \quad [> f_s]$$

Therefore, the tensile stress in the steel is 17,000 lbf/in^2.

The answer is (C).

Why Other Options Are Wrong

(A) This incorrect solution uses the clear span for the span length. The span length should include one-half of the bearing length on each side of the opening.

(B) This incorrect solution uses the depth of the lintel (16 in) for d instead of the effective depth to the reinforcement.

(D) This incorrect solution finds the allowable stress in the steel.

SOLUTION 13

A bolt is classified as a dowel type fastener. The allowable load per fastener for dowel type fasteners is given in NDS Table 11.3.1 as

$$Z' = ZC_D C_M C_t C_g C_\Delta C_{\text{eg}} C_{\text{di}} C_{\text{tn}}$$

The diaphragm factor, C_{di}, and the toe-nail factor, C_{tn}, do not apply. The other factors are given as 1.0. Calculate Z and C_g.

NDS Table 12A gives the nominal lateral load for bolts in single shear. Main-member and side-member thickness are both 1½ in. For a bolt diameter of ½ in, with the load parallel to the grain, and red oak wood,

$$Z = 650 \text{ lbf/bolt}$$

From NDS Supplement Table 1B, the areas of the main member and the side member are

$$A_m = A_s = 8.25 \text{ in}^2$$

The spacing between fasteners, s, is 6.5 in. The group action factor, C_g, is calculated according to NDS Sec. 11.3.6. Note that the group action factors listed in NDS Table 11.3.6A are overly conservative in this case and should not be used, since $D < 1$ in. Use NDS Eq. 11.3-1 instead.

$$\begin{aligned} C_g &= \left(\frac{m(1 - m^{2n})}{n\left((1 + R_{\text{EA}} m^n)(1 + m) - 1 + m^{2n}\right)}\right) \\ &\quad \times \left(\frac{1 + R_{\text{EA}}}{1 - m}\right) \quad \text{[NDS Eq. 11.3-1]} \end{aligned}$$

The number of fasteners in a row is defined in NDS Sec. 11.3.6.2. When $a < b/4$, adjacent rows are considered as one row for the purpose of determining the group action factors.

$$\begin{aligned} b &= \frac{s}{2} = \frac{6.5 \text{ in}}{2} = 3.25 \text{ in} \\ \frac{b}{4} &= \frac{3.25 \text{ in}}{4} = 0.81 \text{ in} \\ a &= 0.75 \text{ in} \quad [< b/4] \end{aligned}$$

Therefore, for the purpose of determining the group action factors, there are two rows of six fasteners in this case.

$$n = 6$$

$$R_{\text{EA}} = \frac{E_s A_s}{E_m A_m} = \frac{\left(1.2\times 10^6\ \frac{\text{lbf}}{\text{in}^2}\right)(8.25\ \text{in}^2)}{\left(1.2\times 10^6\ \frac{\text{lbf}}{\text{in}^2}\right)(8.25\ \text{in}^2)} = 1.0$$

$$\begin{aligned}\gamma &= \left(180{,}000\ \frac{\text{lbf}}{\text{in}}\right)D^{1.5} = \left(180{,}000\ \frac{\text{lbf}}{\text{in}}\right)(0.5)^{1.5}\\ &= 63{,}640\ \text{lbf/in}\end{aligned}$$

$$\begin{aligned}u &= 1+\gamma\left(\frac{s}{2}\right)\left(\frac{1}{E_m A_m}+\frac{1}{E_s A_s}\right)\\ &= 1+\left(63{,}640\ \frac{\text{lbf}}{\text{in}}\right)\left(\frac{3.25\ \text{in}}{2}\right)\\ &\quad\times\left(\frac{1}{\left(1.2\times 10^6\ \frac{\text{lbf}}{\text{in}^2}\right)(8.25\ \text{in}^2)} + \frac{1}{\left(1.2\times 10^6\ \frac{\text{lbf}}{\text{in}^2}\right)(8.25\ \text{in}^2)}\right)\\ &= 1.02\end{aligned}$$

$$m = u-\sqrt{u^2-1} = 1.02-\sqrt{(1.02)^2-1} = 0.82$$

$$\begin{aligned}C_g &= \left(\frac{m(1-m^{2n})}{n\big((1+R_{\text{EA}}m^n)(1+m)-1+m^{2n}\big)}\right)\left(\frac{1+R_{\text{EA}}}{1-m}\right)\\ &= \left(\frac{(0.82)\big(1-(0.82)^{(2)(6)}\big)}{(6)\left(\begin{array}{c}\big(1+(1.0)(0.82)^6\big)(1+0.82)\\ -1+(0.82)^{(2)(6)}\end{array}\right)}\right)\left(\frac{1+1.0}{1-0.82}\right)\\ &= 0.94\end{aligned}$$

The capacity of the connection is

$$\begin{aligned}Z' &= ZC_D C_M C_t C_g C_\Delta\\ &= \left(650\ \frac{\text{lbf}}{\text{bolt}}\right)(12\ \text{bolts})(1.0)(1.0)(1.0)(0.94)(1.0)\\ &= 7332\ \text{lbf}\quad (7300\ \text{lbf})\end{aligned}$$

The answer is (C).

Why Other Options Are Wrong

(A) This incorrect solution finds the capacity per bolt, not the total capacity of the connection.

(B) This incorrect solution uses the group action factors from NDS Table 11.3.6A. This table is overly conservative in this case, because the diameter is less than 1 in and the spacing is less than 4 in, and the footnote to the table indicates that the tabulated values are conservative if even one of the variables is different from those used in the tabulated values.

(D) This incorrect solution does not properly determine the number of fasteners in a row. In this solution, four rows of three fasteners are used. For the purposes of determining the group action factors, NDS Sec. 11.3.6.2 should be used to determine the number of fasteners in a row.

SOLUTION 14

TMS 402 Sec. 7.3.2.6 contains requirements for special reinforced masonry shear walls. Since the problem requires only the minimum seismic reinforcement, only TMS 402 Sec. 7.3.2.6 and Sec. 7.3.2.3.1 apply.

For masonry laid in running bond, TMS 402 Sec. 7.3.2.6 limits the maximum spacing of vertical and horizontal reinforcement to the smallest of $L/3$, $H/3$, or 48 in.

$$s_{\max} = \frac{L}{3} = \frac{(20\ \text{ft})\left(12\ \frac{\text{in}}{\text{ft}}\right)}{3} = 80\ \text{in}$$

$$\begin{aligned}s_{\max} &= \frac{H}{3} = \frac{(10\ \text{ft})\left(12\ \frac{\text{in}}{\text{ft}}\right)}{3} = 40\ \text{in}\quad \text{[controls]}\\ s_{\max} &= 48\ \text{in}\end{aligned}$$

For masonry laid in running bond, TMS 402 Sec. 7.3.2.6.(c).1 specifies that the minimum cross-sectional area of reinforcement in each direction must be at least 0.0007 times the gross cross-sectional area of the wall, A_g, calculated using specified dimensions. The specified thickness of a nominal 12 in concrete masonry wall is 11.63 in.

$$\begin{aligned}\rho_{v,\min} &= 0.0007\\ \rho_{h,\min} &= 0.0007\end{aligned}$$

Since the maximum spacing is limited to 40 in, try one no. 5 bar in bond beams spaced at 32 in on center vertically.

$$\begin{aligned}\rho_h &= \frac{0.31\ \text{in}^2}{(11.63\ \text{in})(32\ \text{in})}\\ &= 0.00083\quad [> \rho_{h,\text{min}},\ \text{OK}]\end{aligned}$$

TMS 402 Sec. 7.3.2.3.1 minimum reinforcement requirements specify that the minimum horizontal reinforcement must be at least 0.2 in^2 of bond beam reinforcement spaced no more than 120 in on center vertically, or two W1.7 joint reinforcing wires with a maximum spacing of 16 in on center vertically, both of which are less than the proposed bars.

For now, use one no. 5 bar in bond beams spaced at 32 in on center vertically.

Determine the minimum required vertical reinforcement. Try no. 5 reinforcing bars spaced at 24 in on center.

$$\begin{aligned}\rho_v &= \frac{0.31\ \text{in}^2}{(11.63\ \text{in})(24\ \text{in})}\\ &= 0.0011\quad [> \rho_{v,\text{min}},\ \text{OK}]\end{aligned}$$

Number 5 reinforcing bars spaced at 24 in on center are more than adequate for the minimum requirement.

Check the requirements of TMS 402 Sec. 7.3.2.6.(c).

$$\begin{aligned}\rho_v + \rho_h &\geq 0.002\\ \rho_v + \rho_h &= 0.00083 + 0.0011\\ &= 0.0019\quad [\text{no good}]\end{aligned}$$

By inspection, one no. 5 bar in bond beams spaced at 24 in on center and one no. 5 bar vertically at 24 in on center will meet the minimum reinforcing requirements.

The answer is (B).

Why Other Options Are Wrong

(A) This incorrect solution only calculates the minimum horizontal reinforcement. Minimum shear reinforcement is required in both directions according to TMS 402 Sec. 7.3.2.6.

(C) This solution incorrectly omits the check for the maximum horizontal spacing required by TMS 402 Sec. 7.3.2.6.

(D) This incorrect solution bases the minimum required horizontal and vertical reinforcement on $0.002A_g$. This is the minimum required for the sum of the horizontal and vertical reinforcements, not the individual amounts.

SOLUTION 15

Determine the eccentricity of the load. If the resultant falls outside the column flanges, the anchor bolts must resist the resulting tension.

$$e = \frac{M}{P} = \frac{(200\ \text{ft-kips})\left(12\ \frac{\text{in}}{\text{ft}}\right)}{320\ \text{kips}} = 7.5\ \text{in}$$

For a W14 × 109 column,

$$\begin{aligned}d &= 14.3\ \text{in}\\ \frac{d}{2} &= \frac{14.3\ \text{in}}{2} = 7.15\ \text{in}\\ t_f &= 0.860\ \text{in}\\ \frac{t_f}{2} &= \frac{0.860\ \text{in}}{2} = 0.430\ \text{in}\\ \frac{d}{2} - \frac{t_f}{2} &= 7.15\ \text{in} - 0.430\ \text{in} = 6.72\ \text{in}\end{aligned}$$

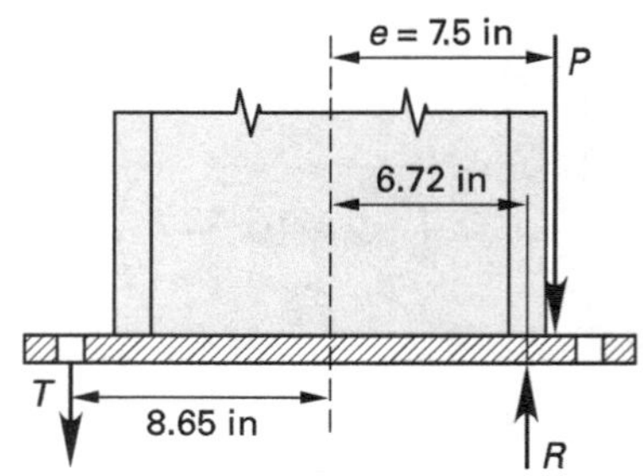

The eccentricity is outside the column flange. Therefore, the bolt must resist uplift. Assume the resultant of the compression forces is located at the center of the column flange. Take moments about this point.

$$\begin{aligned}\sum M_R = 0 &= P\left(e - \left(\frac{d}{2} - \frac{t_f}{2}\right)\right)\\ &\quad - T\left(\left(\frac{d}{2} + 1.5\ \text{in}\right) + \left(\frac{d}{2} - \frac{t_f}{2}\right)\right)\\ 0 &= (320\ \text{kips})(7.5\ \text{in} - 6.72\ \text{in})\\ &\quad - T(8.65\ \text{in} + 6.72\ \text{in})\\ T &= 16.2\ \text{kips}\end{aligned}$$

Calculate the size of bolt required. From Table 7-2 of the AISC *Steel Construction Manual*, determine that a 1 in diameter bolt has an allowable tensile load of 17.7 kips.

The answer is (D).

Why Other Options Are Wrong

(A) This incorrect solution does not consider the moment in the design. For a base plate without any uplift, a $^5/_8$ in diameter bolt is adequate.

(B) This incorrect solution uses the distance to the edge of the column flange instead of the distance to the center of the flange when calculating the tension on the bolt.

(C) This incorrect solution uses the LRFD instead of the ASD column value in the bolt Available Tensile Strength table.

SOLUTION 16

When slenderness must be considered in the design of compression members, the magnified moment procedure can be used if a more refined analysis is not performed.

ACI 318 Sec. 6.6.4.5 contains the provisions for magnified moments in nonsway frames. According to ACI 318 Sec. 6.2.5, if $kl_u/r \leq 34 - 12(M_1/M_2)$ and is no more than 40 for columns braced against sidesway, slenderness can be ignored.

The unsupported length of a compression member is taken as the clear distance between floor slabs.

$$\begin{aligned} l_u &= (13.0\ \text{ft})\left(12\ \frac{\text{in}}{\text{ft}}\right) - 6\ \text{in} \\ &= 150\ \text{in} \end{aligned}$$

For a 12 in diameter column,

$$\begin{aligned} I_g &= \frac{\pi d^4}{64} \\ &= \frac{\pi(12\ \text{in})^4}{64} \\ &= 1018\ \text{in}^4 \\ \frac{kl_u}{r} &= 50 \quad [\text{given}] \\ M_1 &= -100\ \text{ft-kips} \quad \left[\begin{array}{c}\text{for columns bent in} \\ \text{double curvature}\end{array}\right] \\ M_2 &= 100\ \text{ft-kips} \end{aligned}$$

$$\begin{aligned} (34 - 12)\left(\frac{M_1}{M_2}\right) &= 34 - (12)\left(\frac{-100\ \text{ft-kips}}{100\ \text{ft-kips}}\right) \\ &= 46 \quad [\text{so } 40 < kl_u/r] \end{aligned}$$

Therefore, slenderness must be considered, and magnified moments can be used.

The critical buckling load is

$$P_c = \frac{\pi^2(EI)_{\text{eff}}}{(kl_u)^2} \quad [\text{ACI 318 Eq. 6.6.4.4.2}]$$

$$(EI)_{\text{eff}} = \frac{0.4E_cI_g}{1+\beta_{\text{dns}}} \quad [\text{ACI 318 Eq. 6.6.4.4.4a}]$$

According to ACI 318 Sec. R6.6.4.4.4, ACI 318 Eq. 6.6.4.4.4a can be simplified as

$$\begin{aligned} (EI)_{\text{eff}} &= 0.25E_cI_g \\ &= (0.25)\left(3.6\times 10^6\ \frac{\text{lbf}}{\text{in}^2}\right)(1018\ \text{in}^4) \\ &= 9.16\times 10^8\ \text{lbf-in}^2 \end{aligned}$$

$$\begin{aligned} P_c &= \frac{\pi^2(EI)_{\text{eff}}}{(kl_u)^2} \\ &= \frac{\pi^2\left(\dfrac{9.16\times 10^8\ \text{lbf-in}^2}{1000\ \dfrac{\text{lbf}}{\text{kip}}}\right)}{\big((1.0)(150\ \text{in})\big)^2} \\ &= 401.8\ \text{kips} \quad (400\ \text{kips}) \end{aligned}$$

The answer is (A).

Why Other Options Are Wrong

(B) This incorrect solution calculates the gross moment of inertia for a 12 in square column.

(C) This incorrect solution uses E_cI_g instead of $(EI)_{\text{eff}}$.

(D) This incorrect solution uses kl/r instead of kl_u in the equation for P_c.

SOLUTION 17

When slenderness must be considered in the design of compression members, the magnified moment procedure can be used if a more refined analysis is not performed.

ACI 318 Sec. 6.6.4.5 contains the provisions for magnified moments in nonsway frames. According to ACI 318 Sec. 6.2.5, if $kl_u/r \leq 34 - 12(M_1/M_2)$ and is no more than 40 for columns braced against sidesway, slenderness can be ignored.

The unsupported length of a compression member is taken as the clear distance between floor slabs.

$$l_u = (13.0\ \text{ft})\left(12\ \frac{\text{in}}{\text{ft}}\right) - 6\ \text{in} = 150\ \text{in}$$

Design & Details of Structures

For a 12 in diameter column,

$$I_g = \frac{\pi d^4}{64} = \frac{\pi(12 \text{ in})^4}{64} = 1018 \text{ in}^4$$

$$\frac{kl_u}{r} = 50 \quad \text{[given]}$$

$$M_1 = -100 \text{ ft-kips} \quad \left[\begin{array}{c}\text{for columns bent in}\\ \text{double curvature}\end{array}\right]$$

$$M_2 = 100 \text{ ft-kips}$$

$$(34-12)\left(\frac{M_1}{M_2}\right) = 34-(12)\left(\frac{-100 \text{ ft-kips}}{100 \text{ ft-kips}}\right) = 46 \quad [\text{so } 40 < kl_u/r]$$

Therefore, slenderness must be considered, and magnified moments can be used.

The magnified moment for non-sway columns is given by ACI 318 Eq. 6.6.4.5.1.

$$M_c = \delta_{ns} M_2$$

$$\delta_{ns} = \frac{C_m}{1 - \dfrac{P_u}{0.75P_c}} \geq 1.0 \quad \text{[ACI 318 Eq. 6.6.4.5.2]}$$

The critical buckling load, P_c, is given as 405 kips.

$$C_m = 0.6 + 0.4\left(\frac{M_1}{M_2}\right) \quad \text{[ACI 318 Eq. 6.6.4.5.3a]}$$

$$= 0.6 + (0.4)\left(\frac{-100 \text{ ft-kips}}{100 \text{ ft-kips}}\right) = 0.2$$

$$\delta_{ns} = \frac{C_m}{1 - \dfrac{P_u}{0.75P_c}} = \frac{0.2}{1 - \dfrac{300 \text{ kips}}{(0.75)(405 \text{ kips})}} = 16.2 \quad [\geq 1.0, \text{ OK}]$$

$$M_c = \delta_{ns}M_2 = (16.2)(100 \text{ ft-kips}) = 1620 \text{ ft-kips} \quad (1600 \text{ ft-kips})$$

Check $M_2 > M_{2,min}$ (ACI 318 Eq. 6.2.5b).

$$M_{2,min} = P_u(0.6 + 0.03h)$$

$$= (300 \text{ kips})\left(\frac{0.6 \text{ in} + (0.03)(12 \text{ in})}{12 \dfrac{\text{in}}{\text{ft}}}\right) = 24 \text{ ft-kips} \quad [< 100 \text{ ft-kips, OK}]$$

The answer is (B).

Why Other Options Are Wrong

(A) This incorrect solution uses M_1 instead of M_2 in the calculation of the critical moment.

(C) This incorrect solution uses a positive sign for the M_1/M_2 ratio.

(D) This incorrect solution makes a conversion error in the calculation of $M_{2,min}$ that controls the answer.

SOLUTION 18

Section R6.2.5 of ACI 318 contains a nomograph to determine the effective length factor, k, knowing the relative stiffness parameters.

$$\Psi = \frac{\sum\left(\dfrac{EI}{l}\right)_{\text{column}}}{\sum\left(\dfrac{EI}{l}\right)_{\text{beam}}} \quad \text{[ACI 318 Fig. R6.2.5]}$$

Since the modulus of elasticity is the same throughout,

$$\Psi = \frac{\sum\left(\dfrac{I_c}{l_c}\right)_{\text{column}}}{\sum\left(\dfrac{I_b}{l_b}\right)_{\text{beam}}}$$

$$I_c = 8800 \text{ in}^4$$

$$I_b = 10{,}000 \text{ in}^4$$

Adjust for the moment of inertia for cracking and creep [ACI 318 Sec. 6.6.3.1.1].

$$I_{c,e} = 0.70I_c = (0.70)(8800 \text{ in}^4) = 6160 \text{ in}^4$$

$$I_{b,e} = 0.35I_b = (0.35)(10{,}000 \text{ in}^4) = 3500 \text{ in}^4$$

$$\Psi_{top} = \frac{\sum\left(\dfrac{I_c}{l_c}\right)_{\text{column}}}{\sum\left(\dfrac{I_b}{l_b}\right)_{\text{beam}}} = \frac{\dfrac{6160 \text{ in}^4}{13 \text{ ft}} + \dfrac{6160 \text{ in}^4}{13 \text{ ft}}}{\dfrac{3500 \text{ in}^4}{20 \text{ ft}} + \dfrac{3500 \text{ in}^4}{20 \text{ ft}}} = 2.71$$

$$\Psi_{bot} = \frac{\sum\left(\dfrac{I_c}{l_c}\right)_{\text{column}}}{\sum\left(\dfrac{I_b}{l_b}\right)_{\text{beam}}} = \frac{\dfrac{6160 \text{ in}^4}{13 \text{ ft}} + \dfrac{6160 \text{ in}^4}{18 \text{ ft}}}{\dfrac{3500 \text{ in}^4}{20 \text{ ft}} + \dfrac{3500 \text{ in}^4}{20 \text{ ft}}} = 2.33$$

Using the nomograph in ACI 318 Fig. R6.2.5(b) for unbraced frames, the effective length factor in the east/west direction is 1.71.

The effective length of the interior second-story column in the east/west direction is

$$kl_u = (1.71)(12.5 \text{ ft}) = 21.4 \text{ ft} \quad (21 \text{ ft})$$

The answer is (C).

Why Other Options Are Wrong

(A) In this incorrect solution, the relative stiffness parameter is inverted, and the nomograph for braced frames is used.

(B) This incorrect solution does not adjust the moment of inertia for cracking and creep.

(D) This incorrect solution mistakenly uses 13 ft instead of 18 ft for the unbraced column length on the first story.

SOLUTION 19

The allowable axial compressive force is given in TMS 402 Sec. 8.3.4.2.1 for columns with an h/r not greater than 99. Use TMS 402 Eq. 8-21.

$$P_a = (0.25f'_m A_n + 0.65A_{st}F_s)\left(1 - \left(\frac{h}{140r}\right)^2\right)$$

The section properties of the column are based on the actual column dimensions. For a 20 in brick masonry column, the actual dimensions are

$$b = t = 19.625 \text{ in}$$

A no. 8 bar has an area of 0.79 in^2 [ACI 318 App. E]. The area of laterally tied longitudinal reinforcement is

$$\begin{aligned} A_{st} &= \text{four no. 8 bars} = (4)(0.79 \text{ in}^2) \\ &= 3.16 \text{ in}^2 \end{aligned}$$

The net cross-sectional area of the column is

$$\begin{aligned} A_n &= bt - A_{st} = (19.625 \text{ in})(19.625 \text{ in}) - 3.16 \text{ in}^2 \\ &= 382 \text{ in}^2 \end{aligned}$$

The allowable tensile stress for grade 60 reinforcing bars is 32,000 lbf/in^2 (TMS 402 Sec. 8.3.3.1).

$$\begin{aligned} P_a &= (0.25f'_m A_n + 0.65A_{st}F_s)\left(1 - \left(\frac{h}{140r}\right)^2\right) \\ &= \frac{\left(\begin{array}{c}(0.25)\left(3500 \ \frac{\text{lbf}}{\text{in}^2}\right)(382 \text{ in}^2) \\ + (0.65)(3.16 \text{ in}^2)\left(32{,}000 \ \frac{\text{lbf}}{\text{in}^2}\right)\end{array}\right)\left(1 - \left(\frac{72}{140}\right)^2\right)}{1000 \ \frac{\text{lbf}}{\text{kip}}} \\ &= 294 \text{ kips} \end{aligned}$$

The answer is (B).

Why Other Options Are Wrong

(A) This incorrect solution uses an allowable tensile stress of 24,000 lbf for grade 60 steel.

(C) This incorrect solution uses the nominal area of the column (20 in × 20 in) instead of the actual net area.

(D) This incorrect solution uses 1.0 in^2 as the area of a no. 8 bar. A no. 8 bar has a 1.0 in diameter.

SOLUTION 20

The nominal shear strength of concrete for footings with two-way action is given in ACI Sec. 22.6.5.2 as the smallest of

(a) $V_c = 4\lambda\sqrt{f'_c}\, b_o d$

(b) $V_c = \left(2 + \dfrac{4}{\beta}\right)\lambda\sqrt{f'_c}\, b_o d$

(c) $V_c = \left(\dfrac{a_s d}{b_o} + 2\right)\lambda\sqrt{f'_c}\, b_o d$

For a round column, β is 1.0.

ACI 318 Sec. 20.6.1.3.1 requires cover for reinforcement in footings of 3 in. Knowing the diameter of a no. 7 bar is 0.875 in, the depth to the centerline of the layers of reinforcement, d, is

$$\begin{aligned} d &= 24 \text{ in} - 3 \text{ in} - 0.875 \text{ in} \\ &= 20.13 \text{ in} \end{aligned}$$

Design & Details of Structures

The perimeter of the critical section, b_o, is taken at a distance of $d/2$ from the face of the column.

$$b_o = \pi D = \pi(12\text{ in} + 20.13\text{ in}) = 100.94\text{ in} \quad (101\text{ in})$$

$\alpha_s = 40$ for interior columns [ACI 318 Sec. 22.6.5.3]

$\lambda = 1.0$ for normalweight concrete

(a)

$$V_c = 4\lambda\sqrt{f'_c}\,b_o d = \frac{(4)(1.0)\left(\sqrt{3000\ \frac{\text{lbf}}{\text{in}^2}}\right)(101\text{ in})(20.13\text{ in})}{1000\ \frac{\text{lbf}}{\text{kip}}} = 445\text{ kips} \quad \text{[controls]}$$

(b)

$$V_c = \left(2+\frac{4}{\beta}\right)\lambda\sqrt{f'_c}\,b_o d = \frac{\left(2+\frac{4}{1.0}\right)(1.0)\left(\sqrt{3000\ \frac{\text{lbf}}{\text{in}^2}}\right)(101\text{ in})(20.13\text{ in})}{1000\ \frac{\text{lbf}}{\text{kip}}} = 668\text{ kips}$$

(c)

$$V_c = \left(\frac{\alpha_s d}{b_o}+2\right)\lambda\sqrt{f'_c}\,b_o d = \frac{\left(\frac{(40)(20.13\text{ in})}{101\text{ in}}+2\right)(1.0)\left(\sqrt{3000\ \frac{\text{lbf}}{\text{in}^2}}\right)(101\text{ in}) \times (20.13\text{ in})}{1000\ \frac{\text{lbf}}{\text{kip}}} = 1111\text{ kips}$$

Neglecting the shear capacity of the reinforcement, the design punching shear capacity is

$$\phi V_n = \phi V_c = (0.75)(445\text{ kips}) = 334\text{ kips} \quad (330\text{ kips})$$

The answer is (B).

Why Other Options Are Wrong

(A) This incorrect solution calculates the diameter of the critical section as equal to the column diameter plus $d/2$ instead of the column diameter plus $(2)(d/2)$.

(C) This incorrect solution calculates the distance to reinforcement, d, to the bottom of the bottom layer of reinforcement rather than to the center of the reinforcing layers.

(D) This incorrect solution calculates the nominal shear strength instead of the design shear strength.

SOLUTION 21

The beam-to-post connection shown imparts a lateral load on the wood screws. Because two wood members are joined in this connection, the connection is subjected to single shear. Appendix I of the NDS illustrates connections in single and double shear. Use NDS Chap. 11 and Chap. 12 and NDS Table 12L to determine the allowable lateral load on the connection.

The allowable lateral design value for a single connector is given in NDS Table 11.3.1 as

$$Z' = ZC_DC_MC_tC_gC_\Delta C_{eg}C_{di}C_{tn}$$

In this case, C_D is 1.0 for dead load plus live load, since the load duration factor for the shortest duration load applies (NDS Sec. 2.3.2 and App. B). C_M is 0.7, since the moisture content of this deck built in a humid climate will exceed 19% (NDS Table 11.3.3). Values for C_g and C_Δ depend on the diameter of the screw, D. Since $D <$ 0.25 in for a 12-gage screw ($D = 0.216$ in), C_g and C_Δ equal 1.0 (NDS Sec. 11.3.6.1 and Sec. 12.5.1). C_{eg}, C_{di}, and C_{tn} do not apply. C_t is 1.0 since it is a cool climate.

Therefore,

$$Z' = ZC_DC_MC_tC_gC_\Delta = Z(1.0)(0.7)(1.0)(1.0)(1.0)$$

The cut-thread wood screw design values are given in NDS Table 12L. First, determine the side-member thickness. The side member is the 2×8 beam. Using Table 1B of the NDS Supplement, the side-member thickness is 1.5 in. Next, find the column for southern pine and the row for a 12-gage wood screw. The tabulated design value is

$$Z = 161\text{ lbf}$$

$$Z' = ZC_DC_MC_tC_gC_\Delta = (161\text{ lbf})(1.0)(0.7)(1.0)(1.0)(1.0) = 113\text{ lbf} \quad \text{[per screw]}$$

The capacity of the connection is

$$Z'_{\text{total}} = (5\text{ screws})\left(113\ \frac{\text{lbf}}{\text{screw}}\right) = 565\text{ lbf} \quad (570\text{ lbf})$$

The answer is (C).

Why Other Options Are Wrong

(A) This incorrect solution misreads NDS Table 12L and uses the value for a 10-gage screw ($Z = 128$ lbf) instead of the 12-gage screw value.

(B) This incorrect solution uses the wrong load duration factor. The load duration factor for the shortest-duration load applies. This solution uses the smallest load duration factor, $C_D = 0.9$.

(D) This incorrect solution fails to apply the adjustment factors to the tabulated design value to determine the allowable design value.

SOLUTION 22

Section 11.5.4 of ACI 318 covers design of walls under in-plane shear loads. The wall supports a uniform gravity load, w_u. The concrete shear strength is

$$V_c = 2\lambda\sqrt{f'_c}\,hd \quad \text{[ACI 318 Sec. 11.5.4.5]}$$

For normalweight concrete, $\lambda = 1.0$.

$$\begin{aligned} V_c &= (2)(1.0)\sqrt{4000\ \frac{\text{lbf}}{\text{in}^2}}\,\frac{(12\ \text{in})(110\ \text{in})}{1000\ \frac{\text{lbf}}{\text{kip}}} \\ &= 167\ \text{kips} \\ V_n &= V_c + V_s = 167\ \text{kips} + 700\ \text{kips} \\ &= 867\ \text{kips} \end{aligned}$$

The nominal shear strength of the wall is limited by ACI 318 Sec. 11.5.4.3 to

$$\begin{aligned} V_n &= 10\sqrt{f'_c}\,hd \\ &= 10\sqrt{4000\ \frac{\text{lbf}}{\text{in}^2}}\,\frac{(12\ \text{in})(110\ \text{in})}{1000\ \frac{\text{lbf}}{\text{kip}}} \\ &= 834.8\ \text{kips} \quad \text{[controls]} \end{aligned}$$

Therefore, the nominal shear strength of the wall is 834.8 kips (830 kips).

The answer is (C).

Why Other Options Are Wrong

(A) This incorrect solution calculates the shear strength of the concrete instead of the shear strength of the wall.

(B) This incorrect solution calculates the factored shear strength of the wall.

(D) This incorrect solution ignores the limit on nominal shear strength found in ACI 318 Sec. 11.5.4.3.

SOLUTION 23

The design axial strength of a wall using the simplified method is given in ACI Sec. 11.5.3.

$$\phi P_n = 0.55\phi f'_c A_g\left(1 - \left(\frac{kl_c}{32h}\right)^2\right) \quad \text{[ACI 318 Eq. 11.5.3.1]}$$

Calculate the factored axial load, and set it equal to the design axial strength.

$$P_u = w_u l_w = \left(5\ \frac{\text{kips}}{\text{ft}}\right)(16\ \text{ft}) = 80\ \text{kips}$$

$$\begin{aligned} P_u = \phi P_n &= 0.55\phi f'_c A_g\left(1 - \left(\frac{kl_c}{32h}\right)^2\right) \\ &= 0.55\phi f'_c(hl_w)\left(1 - \left(\frac{kl_c}{32h}\right)^2\right) \end{aligned}$$

$$\phi = 0.65 \quad \text{[ACI 318 Sec. 11.5.3.3 and Sec. 21.2.2]}$$

$$\begin{aligned} 80\ \text{kips} &= (0.55)(0.65)\left(4000\ \frac{\text{lbf}}{\text{in}^2}\right)\left(1000\ \frac{\text{lbf}}{\text{kip}}\right) \\ &\quad \times (12\ \text{in})(16\ \text{ft})\left(12\ \frac{\text{in}}{\text{ft}}\right) \\ &\quad \times \left(1 - \left(\frac{(1.0)l_c}{(32)(12\ \text{in})}\right)^2\right) \\ &= (3295\ \text{kips})\left(1 - \frac{l_c^2}{147{,}456\ \text{in}^2}\right) \end{aligned}$$

$$0.024282 = 1 - \frac{l_c^2}{147{,}456\ \text{in}^2}$$

$$\frac{l_c^2}{147{,}456\ \text{in}^2} = 0.975718$$

$$\begin{aligned} l_c &= (379.3\ \text{in})\left(12\ \frac{\text{in}}{\text{ft}}\right) \\ &= 31.6\ \text{ft} \end{aligned}$$

Design & Details of Structures

Check the minimum thickness requirements found in ACI 318 Sec. 11.3.1.1. Thickness is based on the shorter of the wall height and wall length. If the wall height is 31.6 ft and the wall length is 16 ft, the minimum thickness is

$$h > \frac{l_w}{25} = \frac{(16 \text{ ft})\left(12 \frac{\text{in}}{\text{ft}}\right)}{25} \geq 4 \text{ in}$$
$$= 7.7 \text{ in} \quad [\text{OK}]$$

The answer is (C).

Why Other Options Are Wrong

(A) This incorrect solution does not properly apply the requirements for minimum thickness found in ACI 318 Sec. 11.3.1.1. The maximum height is incorrectly calculated as the product of 25 times the wall thickness, h. This requirement should be used to check the minimum thickness based upon the lesser of the height or length. In this case, the length, not the height, would be the limiting value.

(B) This incorrect solution calculates the gross area with the length of the wall in feet rather than inches. The units do not work out.

(D) This incorrect solution calculates the uniform axial load, w_u, rather than the total axial load, P_u. The units do not work out.

SOLUTION 24

The allowable bearing capacity of a single pile is

$$Q_a = \frac{Q_u}{F}$$

The ultimate bearing capacity is the sum of the point-bearing capacity, Q_p, and the skin-friction capacity, Q_f.

$$Q_u = Q_p + Q_f$$
$$Q_p = A_p c N_c$$

From a foundation handbook, for driven piles of virtually all conventional dimensions,

$$N_c = 9$$
$$Q_p = 9A_p c$$
$$Q_f = A_s f_s$$
$$f_s = c_A + \sigma_h \tan\delta$$

For saturated clay, the angle of external friction, δ, is 0.

$$Q_f = A_s f_s = A_s c_A$$

The area of the pile tip for a steel H-pile is calculated using the area enclosed by the block perimeter. From AISC Table 1-4, find the block dimensions of an HP12 × 63 to be 11.9 in × 12 in.

$$A_p = db_f = \frac{(11.9 \text{ in})(12 \text{ in})}{\left(12 \frac{\text{in}}{\text{ft}}\right)^2} = 0.99 \text{ ft}^2$$

$$Q_p = 9A_p c = \frac{(9)(0.99 \text{ ft}^2)\left(400 \frac{\text{lbf}}{\text{ft}^2}\right)}{1000 \frac{\text{lbf}}{\text{kip}}} = 3.57 \text{ kips}$$

In an H-pile, the area between the flanges is assumed to fill with soil that moves with the pile. In calculating the skin area, A_s, of an H-pile, the perimeter is the block perimeter of the pile.

$$A_s = pL$$
$$= \left((2)(11.9 \text{ in}) + (2)(12 \text{ in})\right)\left(\frac{1 \text{ ft}}{12 \text{ in}}\right)(100 \text{ ft})$$
$$= 398 \text{ ft}^2$$

$$Q_f = A_s c_A = \frac{(398 \text{ ft}^2)\left(360 \frac{\text{lbf}}{\text{ft}^2}\right)}{1000 \frac{\text{lbf}}{\text{kip}}} = 143.4 \text{ kips}$$

$$Q_u = Q_p + Q_f = 3.57 \text{ kips} + 143.4 \text{ kips} = 147 \text{ kips}$$

With a factor of safety of 3, the allowable load is

$$Q_a = \frac{Q_u}{F} = \frac{147 \text{ kips}}{3} = 48.99 \text{ kips} \quad (49 \text{ kips})$$

The answer is (B).

Why Other Options Are Wrong

(A) This incorrect solution does not include the length of the pile in calculating the skin area. The units do not work out.

(C) This incorrect solution miscalculates the perimeter area of the pile. When determining the pile capacity, the skin area and the point area of an H-pile should be calculated using the block perimeter and block area of the pile. In this wrong solution, the actual pile perimeter is used, as is the actual area of steel section.

(D) This incorrect solution does not include the factor of safety in determining the allowable capacity.

SOLUTION 25

Begin by determining the lateral earth pressure on the wall using the Rankine active pressure, k_a.

The lateral earth pressure due to the soil backfill is

$$p_a = k_a p_v = k_a \gamma H$$

At the bottom of the wall,

$$H = 20 \text{ ft}$$

$$p_a = k_a \gamma H = (0.361)\left(0.110 \ \frac{\text{kip}}{\text{ft}^3}\right)(20 \text{ ft}) = 0.794 \text{ kip/ft}^2$$

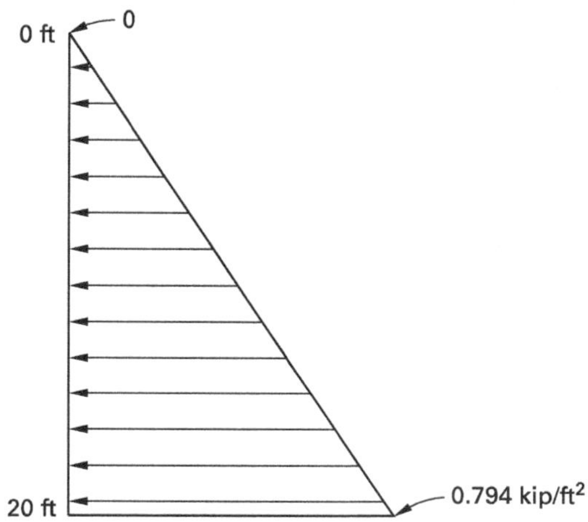

The total active resultant is

$$R_a = \tfrac{1}{2} k_a \gamma H^2 = \tfrac{1}{2} p_a H = \left(\frac{1}{2}\right)\left(0.794 \ \frac{\text{kip}}{\text{ft}^2}\right)(20 \text{ ft}) = 7.94 \text{ kips/ft}$$

Calculate the overturning moment about the toe.

$$M_{\text{overturning}} = R_a y_a = \left(7.94 \ \frac{\text{kips}}{\text{ft}}\right)\left((20 \text{ ft})\left(\frac{1}{3}\right)\right) = 52.9 \text{ ft-kips/ft}$$

Overturning is resisted by the weight of the soil acting vertically on the footing and the weight of the retaining wall. Passive restraint from the soil in front of the footing is only considered if it will always be there. In most cases, it is neglected.

$$M_{\text{resisting}} = \sum W_i x_i$$

$$W_i = \gamma_i A_i$$

Soil:

$$A_{\text{soil}} = LH = \left(15 \text{ ft} - (20 \text{ in})\left(12 \ \frac{\text{in}}{\text{ft}}\right)\right)(20 \text{ ft} - 2 \text{ ft}) = 240 \text{ ft}^2$$

$$W_{\text{soil}} = \gamma_{\text{soil}} A_{\text{soil}} = \left(0.110 \ \frac{\text{kip}}{\text{ft}^3}\right)(240 \text{ ft}^2) = 26.4 \text{ kips/ft}$$

Retaining Wall, Vertical Section:

$$A_{w,v} = LH = (20 \text{ in})\left(12 \ \frac{\text{in}}{\text{ft}}\right)(20 \text{ ft}) = 33.3 \text{ ft}^2$$

$$W_{w,v} = \gamma_w A_{w,v} = \left(0.150 \ \frac{\text{kip}}{\text{ft}^3}\right)(33.3 \text{ ft}^2) = 5.00 \text{ kips/ft}$$

Retaining Wall, Horizontal Section:

$$A_{w,h} = LH = \left(\frac{15 \text{ ft} - 20 \text{ in}}{12 \ \frac{\text{in}}{\text{ft}}}\right)(2 \text{ ft}) = 26.7 \text{ ft}^2$$

$$W_{w,h} = \gamma_w A_w = \left(0.150 \ \frac{\text{kip}}{\text{ft}^3}\right)(26.7 \text{ ft}^2) = 4.01 \text{ kips/ft}$$

The resisting moment about the toe is

$$\begin{aligned} M_{\text{resisting}} &= \sum W_i x_i \\ &= W_{\text{soil}} x_{\text{soil}} + W_{w,v} x_{w,v} + W_{w,h} x_{w,h} \\ &= \left(26.4 \ \frac{\text{kips}}{\text{ft}}\right)\left(\frac{13.33 \text{ ft}}{2} + (20 \text{ in})\left(12 \ \frac{\text{in}}{\text{ft}}\right)\right) \\ &\quad + \left(5.00 \ \frac{\text{kips}}{\text{ft}}\right)\left(\frac{1}{2}\right)(20 \text{ in})\left(12 \ \frac{\text{in}}{\text{ft}}\right) \\ &\quad + \left(4.01 \ \frac{\text{kips}}{\text{ft}}\right)\left(\frac{13.33 \text{ ft}}{2} + (20 \text{ in})\left(12 \ \frac{\text{in}}{\text{ft}}\right)\right) \\ &= 257 \text{ ft-kips/ft} \end{aligned}$$

The factor of safety against overturning is

$$F_{\text{OT}} = \frac{M_{\text{resisting}}}{M_{\text{overturning}}} = \frac{257 \ \frac{\text{ft-kips}}{\text{ft}}}{52.9 \ \frac{\text{ft-kips}}{\text{ft}}} = 4.86$$

The answer is (C).

Why Other Options Are Wrong

(A) This incorrect solution correctly calculates the overturning moment but neglects the weight of the wall in calculating the resisting moment.

(B) This incorrect solution neglects the weight of the horizontal portion of the wall when calculating the resisting moment.

(D) This incorrect solution miscalculates the area of the soil when determining the resisting moment by using the full wall height instead of the soil height.

SOLUTION 26

ACI Sec. 9.7.6 covers the transverse reinforcement of beams. The maximum spacing of the transverse reinforcement is the lesser of the required spacing and the maximum spacing for torsion.

Determine the required transverse reinforcement and spacing. When torsional reinforcement is required, as in this case, ACI Sec. 9.6.4.2 requires that the minimum transverse reinforcement is the greater of

$$\begin{aligned}\frac{A_{v,\min}}{s} &= 0.75\sqrt{f'_c}\left(\frac{b_w}{f_{yt}}\right)\\ &= 0.75\sqrt{4000\ \frac{\text{lbf}}{\text{in}^2}}\left(\frac{35\ \text{in}}{40{,}000\ \frac{\text{lbf}}{\text{in}^2}}\right)\\ &= 0.0415\ \text{in}^2/\text{in}\end{aligned}$$

$$\begin{aligned}\frac{A_{v,\min}}{s} &= \frac{50b_w}{f_{yt}} = \frac{(50)(35\ \text{in})}{40{,}000\ \frac{\text{lbf}}{\text{in}^2}}\\ &= 0.0438\ \text{in}^2/\text{in}\quad\text{[controls]}\end{aligned}$$

However, the minimum transverse reinforcement for torsion is given as 0.05 in^2/in. This is greater than the value of 0.0438 in^2/in determined by ACI Sec. 9.6.4.2, so use $A_{v,\text{req}} = 0.05\ \text{in}^2/\text{in}$.

Number 4 stirrups have a cross-sectional area, A_s, of 0.20 in^2, so the required spacing for no. 4 stirrups is

$$s_{\text{req}} = \frac{A_s}{A_{v,\text{req}}} = \frac{0.20\ \text{in}^2}{0.05\ \frac{\text{in}^2}{\text{in}}} = 4.0\ \text{in}$$

The maximum spacing permitted by ACI Sec. 9.7.6.3.3 for torsion is the lesser of $p_h/8$ or 12 in.

p_h is the perimeter of the area enclosed by the centerline of the transverse reinforcement, or

$$\begin{aligned}p_h &= 2b_t + 2d_t = (2)(31\ \text{in}) + (2)(36\ \text{in})\\ &= 134\ \text{in}\end{aligned}$$

The maximum spacing is the lesser of

$$\begin{aligned}s_{\max} &= \frac{p_h}{8} = \frac{134\ \text{in}}{8}\\ &= 16.75\ \text{in}\\ s_{\max} &\le 12\ \text{in}\quad\text{[controls]}\end{aligned}$$

The maximum spacing of the transverse reinforcement is the lesser of the required spacing, 4.0 in, and the maximum spacing for torsion, 12 in. Use no. 4 stirrups at 4.0 in on center.

The answer is (A).

Why Other Options Are Wrong

(B) This incorrect solution is the spacing for a no. 5 stirrup, not a no. 4 stirrup.

(C) This incorrect solution is the maximum spacing limit determined by ACI Sec. 9.7.6.3.3.

(D) This incorrect solution is the maximum spacing of $p_h/8$.

SOLUTION 27

The nominal shear strength of a fillet weld does not depend on whether the weld is horizontal or vertical, but on the properties of the materials joined and the size and length of the weld.

The nominal strength of a fillet weld, R_n, can be determined from the nominal stress of the weld metal, F_{nw}, and the effective area of the weld, A_{we}, with AISC Eq. J2-3 and Table J2.5.

$$R_n = F_{nw}A_{we}$$

From AISC Table J2.5, ϕ is 0.75, F_{nw} is $0.60F_{\text{EXX}}$ (the weld electrode strength), and A_{we} is the effective length multiplied by the effective throat thickness. Therefore, statements I, II, and IV are true.

The answer is (D).

Why Other Options Are Wrong

(A) This incorrect solution selects the only statement that is false.

(B) This incorrect solution does not include statement IV, which is true.

(C) This incorrect solution includes statement III, which is false. Whether the weld is horizontal or vertical does not affect the shear capacity of the weld.

SOLUTION 28

Section 22.4.2.1 of ACI 318 gives the design axial load strength for compression members. Solve ACI 318 Eq. 22.4.2.2 for the gross area of the concrete.

$$\rho_{g_{\text{max}}} = 0.08 \quad \text{[ACI 318 Sec. 10.6.1.1]}$$

$$\phi = 0.75 \quad \left[\begin{array}{l}\text{for spiral columns,}\\ \text{ACI 318 Sec. 21.2.1}\end{array}\right]$$

$$\begin{aligned} P_u &= 1.2D + 1.6L \\ &= (1.2)(300 \text{ kips}) + (1.6)(350 \text{ kips}) \\ &= 920 \text{ kips} \end{aligned}$$

$$\phi P_{n,\text{max}} = 0.85\phi\left(0.85f_c'(A_g - A_{st}) + f_yA_{st}\right)$$

Replacing A_{st} with $A_g\rho_g$ gives

$$\begin{aligned} P_u = \phi P_{n,\text{max}} &= 0.85\phi\left(0.85f_c'(A_g - A_g\rho_g) + f_yA_g\rho_g\right) \\ &= 0.85\phi A_g\left(0.85f_c'(1-\rho_g) + f_y\rho_g\right) \end{aligned}$$

$$\begin{aligned} A_g &= \frac{P_u}{0.85\phi\left(0.85f_c'(1-\rho_g) + \rho_g f_y\right)} \\ &= \frac{(920 \text{ kips})\left(1000 \frac{\text{lbf}}{\text{kip}}\right)}{(0.85)(0.75)\left(\begin{array}{c}(0.85)\left(4000 \frac{\text{lbf}}{\text{in}^2}\right)(1-0.08) \\ +(0.08)\left(60{,}000 \frac{\text{lbf}}{\text{in}^2}\right)\end{array}\right)} \\ &= 182 \text{ in}^2 \quad (180 \text{ in}^2) \end{aligned}$$

The answer is (B).

Why Other Options Are Wrong

(A) This incorrect solution makes a mathematical error in calculating the cross-sectional area by not multiplying the entire denominator by 0.85ϕ.

$$\begin{aligned} A_g &= \frac{P_u}{0.85\phi 0.85f_c'(1-\rho_g) + \rho_g f_y} \\ &= \frac{(920 \text{ kips})\left(1000 \frac{\text{lbf}}{\text{kip}}\right)}{\begin{array}{c}(0.85)(0.75)(0.85)\left(4000 \frac{\text{lbf}}{\text{in}^2}\right)(1-0.08) \\ +(0.08)\left(60{,}000 \frac{\text{lbf}}{\text{in}^2}\right)\end{array}} \\ &= 135 \text{ in}^2 \quad (140 \text{ in}^2) \end{aligned}$$

(C) This incorrect solution uses $\phi = 0.70$ instead of $\phi = 0.75$.

(D) This incorrect solution uses the adjustment factor for a tied column ($\phi = 0.65$) rather than for a spiral column.

SOLUTION 29

Because the concrete has not gained any of its strength immediately after pouring, the dead loads (including the slab and beam weight) are carried by the steel section alone.

The bending stress in the bottom fibers of the steel beam due to dead load is

$$f = \frac{M}{S}$$

$$\begin{aligned} f_{\text{bot}} = \frac{M_D}{S_x} &= \frac{(80 \text{ ft-kips})\left(12 \frac{\text{in}}{\text{ft}}\right)}{68.4 \text{ in}^3} \\ &= 14 \text{ kips/in}^2 \end{aligned}$$

The answer is (B).

Why Other Options Are Wrong

(A) This incorrect solution uses the section modulus for the transformed section rather than just for the beam. The transformed section modulus would be used in shored construction.

(C) This incorrect solution uses the total moment instead of just the dead load moment.

(D) This incorrect solution uses ft-lbf for moment instead of in-lbf when calculating the stress. The units do not work out.

SOLUTION 30

Frequent variations or reversals of stress can cause fatigue. Appendix 3 of the AISC *Steel Construction Manual* gives requirements for fatigue loading. If the number of loading cycles exceeds 20,000 over the lifetime of the member, fatigue must be considered.

Determine the number of cycles over the lifetime of this structure.

$$\begin{aligned} n_{\text{SR}} &= \left(\frac{\text{cycles}}{\text{day}}\right)(\text{number of years})\left(365 \frac{\text{days}}{\text{yr}}\right) \\ &= \left(200 \frac{\text{cycles}}{\text{day}}\right)(25 \text{ years})\left(365 \frac{\text{days}}{\text{yr}}\right) \\ &= 1{,}825{,}000 \text{ cycles} \quad \left[\begin{array}{c}>20{,}000 \text{ cycles.} \\ \text{Fatigue has to be considered.}\end{array}\right] \end{aligned}$$

Two questions are used to determine whether App. 3 is applicable for use in fatigue design.

1. Is the member subject to normal atmospheric conditions?

2. Is it subject to temperatures less than 300°F?

For this problem, the answer to both questions is "yes." Therefore, App. 3 is applicable, and the maximum stress for unfactored loads shall be $\leq 0.66F_y$. Note that service loads, not factored loads, are used for fatigue design.

From AISC *Steel Construction Manual* Table A-3.1, determine that a bolted end connection is shown in Ex. 2.1 and Ex. 2.2. From AISC Table A-3.1, determine that

stress category	B
constant, C_f	120×10^8
threshold, F_{TH}	16 kips/in^2

The design stress range, F_{SR} (in kips/in^2), is given as

$$F_{SR} = \left(\frac{C_f}{n_{SR}}\right)^{0.333} \quad \text{[AISC Eq. A-3-1]}$$
$$= \left(\frac{120 \times 10^8}{1{,}825{,}000}\right)^{0.333}$$
$$= 18.68 \text{ kips/in}^2 \quad [> F_{TH}]$$

Use F_{SR} equal to 18.68 kips/in^2.

Determine the range of loads.

$$P_{max} = D + L = 20 \text{ kips} + 50 \text{ kips} = 70 \text{ kips (tension)}$$
$$P_{min} = D - L = 20 \text{ kips} - 10 \text{ kips} = 10 \text{ kips (tension)}$$

The allowable tensile stress is

$$F_t = 0.66F_y = (0.66)\left(36 \ \frac{\text{kips}}{\text{in}^2}\right) = 23.8 \text{ kips/in}^2$$

Estimate the channel size.

$$A_{req} = \frac{P_{max}}{F_t} = \frac{70 \text{ kips}}{23.8 \ \frac{\text{kips}}{\text{in}^2}} = 2.94 \text{ in}^2$$

Try a C8 × 11.5.

$$A_{prov} = 3.38 \text{ in}^2$$
$$f_{t,max} = \frac{P_{max}}{A_{prov}} = \frac{70 \text{ kips}}{3.38 \text{ in}^2} = 20.7 \text{ kips/in}^2$$
$$f_{t,min} = \frac{P_{min}}{A_{prov}} = \frac{10 \text{ kips}}{3.38 \text{ in}^2} = 2.96 \text{ kips/in}^2$$

The actual stress range is

$$f_{SR} = 20.7 \ \frac{\text{kips}}{\text{in}^2} - 2.96 \ \frac{\text{kips}}{\text{in}^2} = 17.7 \text{ kips/in}^2$$

The allowable (design) stress range is

$$F_{SR} = 18.68 \text{ kips/in}^2 \quad [>17.7 \text{ kips/in}^2, \text{OK}]$$

Use a C8 × 11.5.

The answer is (C).

Why Other Options Are Wrong

(A) This incorrect solution adds the alternating live loads together instead of treating them as separate loading conditions.

(B) This incorrect solution uses a design stress range of 16 kips/in^2, which is the threshold stress.

(D) This incorrect solution uses the threshold stress range as the actual design stress.

SOLUTION 31

Section I3 of the AISC *Steel Construction Manual* states that the total horizontal shear to be resisted between the point of maximum positive moment and the points of zero moment is the lesser of

$$V' = 0.85f'_cA_c = (0.85)\left(3 \ \frac{\text{kips}}{\text{in}^2}\right)(288 \text{ in}^2) = 734 \text{ kips} \quad \text{[AISC Eq. I3-1a]}$$
$$V' = F_yA_s = \left(50 \ \frac{\text{kips}}{\text{in}^2}\right)(11.8 \text{ in}^2) = 590 \text{ kips} \quad \text{[AISC Eq. I3-1b]}$$

The horizontal shear force that must be carried by the studs is 590 kips.

$$V' = \sum Q_n \quad \text{[AISC Eq. I3-1c]}$$
$$590 \text{ kips} = \sum Q_n$$

The allowable horizontal shear load on a ¾ in × 3½ in headed stud in the strong direction and lightweight concrete is found from AISC Table 3-21.

$$Q_n = 17.1 \text{ kips/stud}$$

The number of studs required between the point of maximum positive moment and the points of zero moment is

$$n = \frac{\sum Q_n}{Q_n} = \frac{590 \text{ kips}}{17.1 \frac{\text{kips}}{\text{stud}}} = 34.5 \text{ studs} \quad \text{[Use 35 studs.]}$$

For a simply supported beam, the maximum positive moment occurs at the midspan of the beam. Therefore, the total number of shear studs required is

$$n_{\text{total}} = 2n = (2)(35 \text{ studs}) = 70 \text{ studs}$$

The answer is (C).

Why Other Options Are Wrong

(A) This incorrect solution only calculates the number of studs required for one-half of the span.

(B) This incorrect solution uses the value for normal-weight concrete in determining the stud capacity.

(D) This incorrect solution uses the concrete shear as the limiting V' instead of the lesser value.

SOLUTION 32

Begin by determining the section properties of the glulam rafters. Table 1D in the NDS Supplement lists the section properties of 8½ in × 27½ in southern pine glulam timber.

$$A = 233.8 \text{ in}^2$$
$$S_x = 1071 \text{ in}^3$$

The allowable bending design value, F'_b, is given in NDS Table 5.3.1 as

$$F'_b = F_{b,xx} C_D C_M C_t C_L C_{\text{fu}} C_I$$

Or,

$$F'_b = F_{b,xx} C_D C_M C_t C_V C_{\text{fu}} C_I$$

The smaller of the two equations is the critical one, because C_V and C_L should not apply simultaneously for glulam timber bending members according to NDS Sec. 5.3.6.

Not all of the adjustment factors are applicable in this case. The load duration factor, C_D, is based on the shortest duration load in a combination of loads. For dead load plus live load, use C_D for live load, 1.0. The wet service factor, C_M, and the temperature factor, C_t, do not apply because the environment is neither wet nor hot. The flat-use factor, C_{fu}, does not apply because the rafters are not used in this way [NDS Sec. 5.3.7]. The curvature factor, C_I, does not apply because the rafters are not tapered.

The allowable bending design value, F'_b, with the applicable adjustment factors is

$$F'_b = F_{b,xx} C_D C_L$$

Or,

$$F'_b = F_{b,xx} C_D C_V$$

The load duration factor, C_D, is based on the shortest duration load in a combination of loads. For the plus live load combination, use C_D for live load, 1.0.

Calculate the beam stability factor, C_L, and the volume factor, C_V. The smaller value will be used to calculate the allowable bending design value.

The beam stability factor is 1.0 when the compression edge of a bending member is supported throughout its length to prevent lateral displacements and when the ends at points of bearing have lateral support to prevent rotation (NDS Sec. 3.3.3.3). The volume factor is found in NDS Sec. 5.3.6. Using NDS Eq. 5.3-1,

$$C_V = \left(\frac{21}{L}\right)^{1/x} \left(\frac{12}{d}\right)^{1/x} \left(\frac{5.125}{b}\right)^{1/x} \leq 1.0$$

L is the length of the bending member between points of zero moment, in feet. x is 20 for southern pine and 10 for all other species.

In this case, L is 15 ft, d is 27.5 in, b is 8.5 in, and x is 20.

$$\begin{aligned} C_V &= \left(\frac{21}{L}\right)^{1/x}\left(\frac{12}{d}\right)^{1/x}\left(\frac{5.125}{b}\right)^{1/x} \\ &= \left(\frac{21}{15\text{ ft}}\right)^{1/20}\left(\frac{12}{27.5\text{ in}}\right)^{1/20}\left(\frac{5.125}{8.5\text{ in}}\right)^{1/20} \\ &= 0.95 \quad [<1.0] \end{aligned}$$

Since C_V is less than C_L, use the equation

$$\begin{aligned} F_b' &= F_{b,xx}C_DC_V \\ &= \left(2000\ \frac{\text{lbf}}{\text{in}^2}\right)(1.0)(0.95) \\ &= 1900\ \text{lbf/in}^2 \end{aligned}$$

The answer is (C).

Why Other Options Are Wrong

(A) This incorrect solution applies a temperature adjustment factor of 0.8 when calculating the allowable bending design value. NDS Commentary states that temperature adjustment factors are not necessary for temperature fluctuations due to solar radiation.

(B) This incorrect solution uses a value of 10 for x in the calculation of the volume adjustment factor.

(D) This incorrect solution neglects to include the volume adjustment factor, C_V, in calculating the allowable bending design value.

SOLUTION 33

The nominal shear strength of a fillet weld is given in Table J2.5 of the AISC *Steel Construction Manual* as

$$F_{nw} = 0.60F_{\text{EXX}}$$

The nominal strength of the weld metal is determined by the electrodes used. E70XX electrodes have a strength of 70 kips/in^2.

The available weld length can be taken as the workable flat dimension or the nominal dimension of the tube less the radius at each end. According to the AISC *Manual*, the outside corner radii are taken as $2.25t_{\text{nom}}$.

$$\begin{aligned} L_w &= 2\big(b-(2)(2.25t_{\text{nom}})\big) \\ &= (2)\big(4\text{ in}-(2)(2.25)(0.375\text{ in})\big) \\ &= 4.63\text{ in} \end{aligned}$$

The effective throat, t_e, of a fillet weld using the SMAW process is $0.707w$. The shear capacity of a weld is given by AISC Eq. J2-3 as

$$\begin{aligned} R_n &= F_{nw}A_{we} = 0.60F_{\text{EXX}}L_wt_e \\ &= (0.60)\left(70\ \frac{\text{kips}}{\text{in}^2}\right)(4.63\text{ in})t_e \end{aligned}$$

The allowable strength, R_n/Ω, must equal or exceed the load on the weld.

$$12{,}000\text{ lbf} = \frac{R_n}{2.00}$$

Solving for the required effective throat thickness,

$$\begin{aligned} t_e &= \frac{(12{,}000\text{ lbf})(2.00)}{\left(42\ \frac{\text{kips}}{\text{in}^2}\right)(4.63\text{ in})\left(1000\ \frac{\text{lbf}}{\text{kip}}\right)} \\ &= 0.123\text{ in} \end{aligned}$$

The nominal weld size is

$$\begin{aligned} w &= \frac{t_e}{0.707} = \frac{0.123\text{ in}}{0.707} \\ &= 0.174\text{ in} \end{aligned}$$

A $\frac{3}{16}$ in fillet weld provides adequate capacity.

Check the minimum weld size required based on the thicknesses of the materials. AISC Table J2.4 specifies the minimum weld size based on the thickness of the thinner part joined.

The thickness of the W10 × 33 flange is found in AISC Part 1.

$$t_f = 0.435\text{ in}$$

The thickness of the HSS4 × 4 × $\frac{3}{8}$ tube is $\frac{3}{8}$ in or 0.375 in. Therefore, the tube thickness controls. The minimum weld size from AISC Table J2.4 is $\frac{3}{16}$ in.

Use a $\frac{3}{16}$ in fillet weld.

The answer is (B).

Why Other Options Are Wrong

(A) This incorrect solution uses the nominal tube length in determining the weld length and does not check the minimum weld size based on the thicknesses of the materials.

(C) This incorrect solution uses the flange thickness of a W10 × 39 column instead of a W10 × 33 column. In this case, the thicknesses of materials controls the size of the weld.

(D) This incorrect solution does not use the same units for the nominal weld strength and the load on the weld when calculating the required weld size.

SOLUTION 34

Part 10 of the AISC *Steel Construction Manual* covers connection design. Using the table for double angle connections, the maximum beam reaction is the lesser of the available strength of the bolt and angle and the available strength of the beam web.

$$\text{L}3\tfrac{1}{2} \times 3\tfrac{1}{2} \times \frac{5}{16} \text{ angles}$$

$$\frac{3}{4} \text{ in diameter A325-N bolts}$$

$$\begin{aligned} L_{eh} &= 1.75 \text{ in} \\ L_{ev} &= 1.5 \text{ in} \\ t_w &= 0.23 \text{ in} \end{aligned}$$

From AISC Table 10-1, the available strength of the bolt and angle is 95.5 kips.

The beam web available strength is 207 kips/in of beam web thickness.

$$\begin{aligned} R_u &= t_w(\text{tabulated value}) \\ &= (0.23 \text{ in})\left(207 \ \frac{\text{kips}}{\text{in}}\right) \\ &= 47.6 \text{ kips} \end{aligned}$$

The beam web available strength controls. The maximum reaction is 47.6 kips (48 kips).

The answer is (C).

Why Other Options Are Wrong

(A) This incorrect solution uses the ASD column in the table for beam web available strength.

(B) This incorrect solution mistakenly uses the value for L_{eh} equal to 1.5 in, instead of 1.75 in, to determine the beam web available strength.

(D) This incorrect solution only considers the bolt and angle available strength and does not check the beam web available strength.

SOLUTION 35

The welds in this case are subject to eccentric loading. The minimum weld size, in sixteenths of an inch, can be determined using LRFD from the equation

$$D_{\text{min}} = \frac{P_u}{\phi C C_1 l}$$

From Table 8-4 of the AISC *Steel Construction Manual*,

$$\begin{aligned} \phi &= 0.75 \\ C_1 &= 1.0 \text{ for E70XX electrodes} \\ k &= 0 \text{ for special case of load not in the plane of the weld group} \end{aligned}$$

The characteristic length of the weld group, l, is given as 8 in, and the eccentricity is given as 2.25 in. To determine the coefficient, C, in AISC Table 8-4, first determine

$$a = \frac{e_x}{l} = \frac{2.25 \text{ in}}{8 \text{ in}} = 0.28$$

By interpolation, C is equal to 3.18 kips/in. The minimum weld size is

$$\begin{aligned} D_{\text{min}} &= \frac{P_u}{\phi C C_1 l} = \frac{60 \text{ kips}}{(0.75)\left(3.18 \ \frac{\text{kips}}{\text{in}}\right)(1.0)(8 \text{ in})} \\ &= 3.14 \end{aligned}$$

The minimum thickness of the weld is

$$t_w = \frac{D_{\text{min}}}{16} = \frac{3.14}{16} \text{ in} = 4/16 \text{ in} \quad (1/4 \text{ in})$$

Check the minimum weld size based on the thickness of the parts joined, using AISC Table J2.4. The flange thickness of a W12 × 50 is 5/8 in, and the thickness of the angle is 1/2 in. The minimum weld size based on the thinner part joined is 3/16 in, which is less than the 1/4 in required. Use a 1/4 in weld.

The answer is (C).

Why Other Options Are Wrong

(A) This incorrect solution uses a weld length of 16 in rather than 8 in.

(B) The weld size is based on shear alone. The eccentricity of the load is ignored in this incorrect solution.

(D) This incorrect solution applies the ϕ factor to the ASD equation for weld thickness.

SOLUTION 36

Section 12.2.3 of the NDS gives the withdrawal values for a single nail. From NDS Table 12.3.3A, the specific gravity of southern pine is 0.55 for $E = 1{,}700{,}000 \text{ lbf/in}^2$.

NDS Table 12P gives the diameter of a 6d box nail as 0.099 in.

From NDS Table 12.2C, the tabulated withdrawal design value for a specific gravity of 0.55 and a nail diameter of 0.099 in is 31 lbf per inch of penetration.

The design withdrawal value is

$$\begin{aligned} C_D &= 1.6 \quad \text{[for wind load combinations]} \\ C_M &= 1.0 \quad \text{[for moisture content} \le 19\%\text{]} \\ C_t &= 1.0 \quad \text{[according to NDS Table 11.3.4]} \\ C_{\text{tn}} &= 1.0 \quad \text{[for non-toe-nailed connections]} \end{aligned}$$

$$\begin{aligned} W' &= WC_DC_MC_tC_{\text{tn}} \\ &= (31 \text{ lbf})(1.6)(1.0)(1.0)(1.0) \\ &= 49.6 \text{ lbf} \end{aligned}$$

Calculate the maximum area per nail.

$$A = \frac{W'}{w} = \frac{49.6 \text{ lbf}}{36 \frac{\text{lbf}}{\text{ft}^2}} = 1.38 \text{ ft}^2$$

If the rafters are spaced at 16 in on center,

$$s_r = \frac{16 \text{ in}}{12 \frac{\text{in}}{\text{ft}}} = 1.33 \text{ ft}$$

The maximum nail spacing is

$$\begin{aligned} s_n &= \frac{A}{s_r} = \left(\frac{1.38 \text{ ft}^2}{1.33 \text{ ft}}\right)\left(12 \frac{\text{in}}{\text{ft}}\right) \\ &= 12.4 \text{ in} \quad (12 \text{ in}) \end{aligned}$$

The answer is (C).

Why Other Options Are Wrong

(A) This incorrect solution neglects the load duration factor in calculating the withdrawal capacity.

(B) This incorrect solution uses a rafter spacing of 1.5 ft instead of 1.33 ft.

(D) This incorrect solution bases the tabulated withdrawal value on the diameter of a 6d common nail (0.113 in) instead of a 6d box nail.

SOLUTION 37

Interaction diagrams, such as those found in the *Structural Engineering Reference Manual* or other sources, provide a means to determine the required area of steel when the loads and section properties of a column are known.

$$\begin{aligned} \frac{\phi P_n}{A_g} &= \frac{P_u}{A_g} = \frac{750 \text{ kips}}{(20 \text{ in})(24 \text{ in})} \\ &= 1.56 \text{ kips/in}^2 \end{aligned}$$

$$\begin{aligned} \frac{\phi M_n}{A_g h} &= \frac{M_u}{A_g h} = \frac{(600 \text{ ft-kips})\left(12 \frac{\text{in}}{\text{ft}}\right)}{(20 \text{ in})(24 \text{ in})(24 \text{ in})} \\ &= 0.63 \text{ kip/in}^2 \end{aligned}$$

From the relevant interaction diagram (in the *Structural Engineering Reference Manual*), determine that ρ_g is 0.03.

$$\begin{aligned} A_s &= \rho_g A_g = (0.03)(20 \text{ in})(24 \text{ in}) \\ &= 14.4 \text{ in}^2 \quad (14 \text{ in}^2) \end{aligned}$$

The answer is (B).

Why Other Options Are Wrong

(A) This incorrect solution uses the interaction diagram for $\gamma = 0.90$.

(C) This incorrect solution uses the interaction diagram for columns with steel distributed evenly on four faces instead of the diagram for steel on only two faces.

(D) This solution incorrectly calculates the moment factor for use with the interaction diagram. The units do not work out.

$$\frac{\phi M_n}{A_g h} = \frac{M_u}{A_g h} = \frac{600 \text{ ft-kips}}{(20 \text{ in})(24 \text{ in})} = 1.25 \text{ ft-kips/in}^2$$

SOLUTION 38

Since the plan is symmetric, the relative stiffness parameter is the same in each direction for a corner column.

$$\begin{aligned} \Psi_{\text{E-W}} &= \Psi_{\text{N-S}} \\ &= \frac{\sum\left(\frac{EI}{l_c}\right)_{\text{column}}}{\sum\left(\frac{EI}{l}\right)_{\text{beam}}} \quad \left[\begin{array}{c}\text{ACI 318} \\ \text{Fig. R6.2.5}\end{array}\right] \end{aligned}$$

Since E is the same throughout,

$$\Psi = \frac{\sum\left(\frac{I}{l_c}\right)_{\text{column}}}{\sum\left(\frac{I}{l}\right)_{\text{beam}}}$$

$$I_c = 8748 \text{ in}^4$$

$$I_b = 10{,}000 \text{ in}^4$$

Adjust for cracking and creep [ACI 318 Sec. 6.6.3.1.1].

$$\begin{aligned} I_{c,e} &= 0.70I_c = (0.70)(8748 \text{ in}^4) \\ &= 6124 \text{ in}^4 \\ I_{b,e} &= 0.35I_b = (0.35)(10{,}000 \text{ in}^4) \\ &= 3500 \text{ in}^4 \end{aligned}$$

$$\begin{aligned} \Psi_{\text{E-W}} &= \frac{\sum\left(\frac{I}{l_c}\right)_{\text{column}}}{\sum\left(\frac{I}{l}\right)_{\text{beam}}} = \frac{\frac{6124 \text{ in}^4}{13 \text{ ft}} + \frac{6124 \text{ in}^4}{18 \text{ ft}}}{\frac{3500 \text{ in}^4}{20 \text{ ft}}} \\ &= 4.64 \quad (4.6) \end{aligned}$$

Because the column is a corner column, there is only one beam in each direction.

The answer is (B).

Why Other Options Are Wrong

(A) This incorrect solution calculates the relative stiffness parameter for an interior column, not a corner column.

(C) This incorrect solution mistakenly uses 13 ft instead of 18 ft for the unbraced column length on the first story.

(D) This incorrect solution does not adjust the moment of inertia of the columns for cracking and creep as recommended in ACI 318 Sec. 6.6.3.1.1.

SOLUTION 39

From AASHTO Table 3.7.3.1-1, the drag coefficient, C_D, on a semicircular-nosed pier is 0.7. Using AASHTO Sec. 3.7.3.1 and Eq. 3.7.3.1-1, determine the longitudinal stream pressure.

$$p = \frac{C_D \text{v}^2}{1000} = \frac{(0.7)\left(10 \ \frac{\text{ft}}{\text{sec}}\right)^2}{1000} = 0.07 \text{ kip/ft}^2$$

From AASHTO Sec. 3.7.3.1, the longitudinal drag force is the product of the stream pressure and the exposed surface area.

$$\begin{aligned} F &= pA \\ &= \left(0.07 \ \frac{\text{kip}}{\text{ft}^2}\right)(10 \text{ ft})(5 \text{ ft}) \\ &= 3.5 \text{ kips} \end{aligned}$$

The answer is (B).

Why Other Options Are Wrong

(A) This incorrect solution did not square the velocity in the drag pressure calculation.

(C) This incorrect solution assumes flow in the wrong direction and uses a drag coefficient of 1.4.

(D) This solution incorrectly uses the length of the pier (30 ft) instead of the width of the pier (5 ft) when determining the drag force.

SOLUTION 40

Determine the factored loads on the columns.

$$\begin{aligned} P_{u1} &= 1.2P_D + 1.6P_L \\ &= (1.2)(116 \text{ kips}) + (1.6)(64 \text{ kips}) \\ &= 242 \text{ kips} \\ P_{u2} &= 1.2P_D + 1.6P_L \\ &= (1.2)(70 \text{ kips}) + (1.6)(32 \text{ kips}) \\ &= 135 \text{ kips} \end{aligned}$$

The ultimate soil pressure is

$$q_u = \frac{P_{u1} + P_{u2}}{BL}$$

The footing width and length are

$$\begin{aligned} B &= 4.5 \text{ ft} \\ L &= 20 \text{ ft} \\ q_u &= \frac{P_{u1} + P_{u2}}{BL} = \frac{242 \text{ kips} + 135 \text{ kips}}{(4.5 \text{ ft})(20 \text{ ft})} \\ &= 4.19 \text{ kips/ft}^2 \end{aligned}$$

The design moment at the face of column 1 in the longitudinal direction is

$$\begin{aligned} M_u &= \frac{q_u B x^2}{2} = \frac{\left(4.19 \ \frac{\text{kips}}{\text{ft}^2}\right)(4.5 \text{ ft})(5 \text{ ft})^2}{2} \\ &= 236 \text{ ft-kips} \end{aligned}$$

The required reinforcement can be determined easily by utilizing design aids such as Graph A.1a in *Design of Concrete Structures* or equations as shown in the *Structural Engineering Reference Manual.*

Using Graph A.1a in *Design of Concrete Structures*, enter the graph knowing

$$f_c' = 3000 \text{ lbf/in}^2$$
$$f_y = 60{,}000 \text{ lbf/in}^2$$
$$\phi = 0.9 \quad [\text{for flexure}]$$

$$R = \frac{M_u}{\phi b d^2} = \frac{(236 \text{ ft-kips})\left(1000 \ \frac{\text{lbf}}{\text{kip}}\right)}{(0.9)(4.5 \text{ ft})(15 \text{ in})^2} = 259 \text{ lbf/in}^2$$

Determine that

$$\rho = 0.00455$$
$$A_s = \rho b d = (0.00455)(4.5 \text{ ft})\left(12 \ \frac{\text{in}}{\text{ft}}\right)(15 \text{ in}) = 3.69 \text{ in}^2$$

Nine no. 6 bars provide 3.96 in^2 of steel.

The answer is (D).

Why Other Options Are Wrong

(A) This incorrect solution calculates the design moment and the area of steel required for a 12 in unit width and not for the entire width.

(B) This incorrect solution assumes that six bars are to be used (instead of no. 6 bars) in determining the area of reinforcement required.

(C) This incorrect solution uses service loads instead of factored loads. Service loads are used to size the footing, but design of the footing is based on factored loads.

SOLUTION 41

The depth to reinforcement, *d*, is one-half the thickness of the wall, and a 12 in concrete masonry unit has a specified thickness of 11.63 in, so

$$d = \frac{t}{2} = \frac{11.63 \text{ in}}{2} = 5.81 \text{ in}$$

A no. 5 reinforcing bar has an area of 0.31 in^2.

TMS 402 Sec. 9.3.5 covers wall design for out-of-plane loads. According to TMS 402 Commentary Sec. 9.3.5.2, the nominal moment capacity of the wall is

$$M_n = \left(\frac{P_u}{\phi} + A_s f_y\right)\left(d - \frac{a}{2}\right)$$

In this equation, *a* is

$$a = \frac{A_s f_y + \frac{P_u}{\phi}}{0.80 f_m' b}$$

The wall is non-loadbearing, and only the out-of-plane capacity is being considered, so the factored axial load, P_u, is zero. From TMS 402 Sec. 9.1.4.4, the strength reduction factor, ϕ, for masonry subject to flexure is 0.90.

$$a = \frac{A_s f_y + \frac{P_u}{\phi}}{0.80 f_m' b} = \frac{(0.31 \text{ in}^2)\left(60{,}000 \ \frac{\text{lbf}}{\text{in}^2}\right) + \frac{0 \text{ lbf}}{0.90}}{(0.80)\left(2000 \ \frac{\text{lbf}}{\text{in}^2}\right)(24 \text{ in})} = 0.4844 \text{ in}$$

The nominal moment capacity of the wall is

$$\begin{aligned} M_n &= \left(\frac{P_u}{\phi} + A_s f_y\right)\left(d - \frac{a}{2}\right) \\ &= \left(\frac{0 \text{ lbf}}{0.90} + (0.31 \text{ in}^2)\left(60{,}000 \ \frac{\text{lbf}}{\text{in}^2}\right)\right) \\ &\quad \times \left(\frac{5.81 \text{ in} - \frac{0.4844 \text{ in}}{2}}{12 \ \frac{\text{in}}{\text{ft}}}\right) \\ &= 8630 \text{ ft-lbf} \end{aligned}$$

The moment capacity of the wall is

$$\begin{aligned} \phi M_n &= (0.90)(8630 \text{ ft-lbf}) \\ &= 7767 \text{ ft-lbf} \quad (7800 \text{ ft-lbf}) \end{aligned}$$

The answer is (B).

Why Other Options Are Wrong

(A) This incorrect solution uses a yield strength, f_y, of 32,000 lbf/in^2 instead of 60,000 lbf/in^2.

(C) This incorrect solution neglects to multiply the nominal moment by the strength reduction factor, ϕ, of 0.9.

(D) This incorrect solution uses a depth to reinforcement corresponding to the steel near to one face of the wall rather than centered in the wall.

SOLUTION 42

The equivalent axial load procedure can be used to design beam-columns.

$$P_e = P_u + M_{ux}m + M_{uy}mU$$

$$M_{ux} = Pe_x = \frac{(600 \text{ kips})(12 \text{ in})}{12 \frac{\text{in}}{\text{ft}}} = 600 \text{ ft-kips}$$

$$M_{uy} = 0 \text{ ft-kips}$$

Use a W14 shape.

$$m = \frac{24 \text{ in}}{d} = \frac{24 \text{ in}}{14 \text{ in-ft}} = 1.71 \text{ ft}^{-1}$$

$$\begin{aligned} P_e &= P_u + M_{ux}m + M_{uy}mU \\ &= 600 \text{ kips} + (600 \text{ ft-kips})\left(1.71 \frac{1}{\text{ft}}\right) + 0 \text{ kips} \\ &= 1626 \text{ kips} \end{aligned}$$

Estimate a column size using AISC *Steel Construction Manual* (AISC *Manual*) Axial Compression Table 4-1. For an unbraced length of 28 ft using LRFD, try a W14 × 193 section. A W14 × 193 section has a capacity of 1550 kips at an unbraced length of 28 ft. Check if the section works using AISC Sec. H1.1.

If $P_r/P_c \geq 0.2$, then use the modified form of AISC Eq. H1-1a, as given in Part 6 of the AISC *Manual*.

$$\frac{P_r}{P_c} = \frac{P_r}{\phi_c P_n} = \frac{600 \text{ kips}}{1550 \text{ kips}} = 0.39 \quad [\geq 0.2]$$

Use AISC Eq. H1-1a.

$$pP_r + b_x M_{rx} + b_y M_{ry} \leq 1.0 \quad \text{[AISC Eq. H1-1a]}$$

From AISC *Manual* Table 6-1, for a W14 × 193 with an effective length of 28 ft,

$$p = 0.647 \times 10^{-3} \text{ kip}^{-1}$$
$$b_x = 0.727 \times 10^{-3} \text{ (ft-kip)}^{-1}$$

$$\begin{aligned} pP_r + b_x M_{rx} + b_y M_{ry} &= \left(0.647 \times 10^{-3} \frac{1}{\text{kip}}\right)(600 \text{ kips}) \\ &\quad + \left(\left(0.727 \times 10^{-3} \frac{1}{\text{ft-kip}}\right) \times (600 \text{ ft-kips})\right) \\ &\quad + 0 \text{ ft-kips} \\ &= 0.82 \quad [\leq 1.0] \end{aligned}$$

A W14 × 193 will work, but it is conservative. Try a smaller section.

Check a W14 × 176. For an unbraced length of 28 ft,

$$\phi_c P_n = 1400 \text{ kips}$$

$$\frac{P_r}{P_c} = \frac{P_r}{\phi_c P_n} = \frac{600 \text{ kips}}{1400 \text{ kips}} = 0.43 \quad [>0.2]$$

Use Eq. H1-1a.

From AISC *Manual* Table 6-1, for a W14 × 176 and an effective length of 28 ft,

$$p = 0.715 \times 10^{-3} \text{ kip}^{-1}$$
$$b_x = 0.814 \times 10^{-3} \text{ (ft-kip)}^{-1}$$

$$\begin{aligned} pP_r + b_x M_{rx} + b_y M_{ry} &= \left(0.715 \times 10^{-3} \frac{1}{\text{kip}}\right)(600 \text{ kips}) \\ &\quad + \left(\left(0.814 \times 10^{-3} \frac{1}{\text{ft-kip}}\right) \times (600 \text{ ft-kips})\right) \\ &\quad + 0 \text{ ft-kips} \\ &= 0.92 \quad [\leq 1.0] \end{aligned}$$

Use a W14 × 176.

The answer is (A).

Why Other Options Are Wrong

(B) This incorrect solution does not perform the necessary number of iterations.

(C) This incorrect solution assumes the load is not factored and uses the ASD values.

(D) This incorrect solution assumes the eccentricity about the wrong axis.

SOLUTION 43

From ACI 318 Sec, 11.6.1, the amount of horizontal shear reinforcement required depends upon whether or not $V_u < \phi V_c/2$.

$$\phi V_c = \phi 2\lambda\sqrt{f_c'}\,hd$$
$$\phi = 0.75 \quad \text{[ACI Sec. 21.2.1]}$$
$$\lambda = 1.0 \quad \text{[for normalweight concrete]}$$
$$h = 10 \text{ in}$$
$$d = 110 \text{ in}$$

$$\phi V_c = \phi 2\lambda\sqrt{f_c'}hd$$

$$= \frac{(0.75)(2)(1.0)\sqrt{6000 \ \frac{\text{lbf}}{\text{in}^2}}(10 \text{ in})(110 \text{ in})}{1000 \ \frac{\text{lbf}}{\text{kip}}}$$

$$= 128 \text{ kips}$$

$$\frac{\phi V_c}{2} = \frac{128 \text{ kips}}{2}$$

$$= 64.0 \text{ kips} \quad [> V_u = 60 \text{ kips}]$$

The in-plane $V_u \leq \phi V_c/2$, so the minimum reinforcement is determined according to ACI Table 11.6.1. In this table, the ratio of horizontal shear reinforcement area to the gross area of vertical section, ρ_t, is based on the size and yield strength of the reinforcement. In this case, the largest bar size is no. 5, and the yield strength of the reinforcing steel is 60,000 lbf/in^2, so from ACI Table 11.6.1, ρ_t is 0.0020.

The minimum horizontal reinforcement is

$$A_v = \rho_t A_g = \rho_t h h_w$$

$$= (0.0020)(10 \text{ in})\left((30 \text{ ft})\left(12 \ \frac{\text{in}}{\text{ft}}\right)\right)$$

$$= 7.20 \text{ in}^2$$

Check the spacing limits in ACI 318 Sec. 11.7.3.1. The maximum spacing is the smallest of

$$s_2 \leq \frac{l_w}{5} = \frac{138 \text{ in}}{5} = 27 \text{ in}$$

$$s_2 \leq 3h = (3)(10 \text{ in}) = 30 \text{ in}$$

$$s_2 \leq 18 \text{ in} \quad [\text{controls}]$$

Try no. 5 bars at 18 in on center.

$$A_{\text{prov}} = \left(\frac{0.31 \text{ in}^2}{18 \text{ in}}\right)\left((30 \text{ ft})\left(12 \ \frac{\text{in}}{\text{ft}}\right)\right)$$

$$= 6.20 \text{ in}^2 \quad [< 7.20 \text{ in}^2, \text{ no good}]$$

More reinforcement is needed. Try no. 5 bars at 12 in on center.

$$A_{\text{prov}} = \left(\frac{0.31 \text{ in}^2}{12 \text{ in}}\right)\left((30 \text{ ft})\left(12 \ \frac{\text{in}}{\text{ft}}\right)\right)$$

$$= 9.30 \text{ in}^2 \quad [> 7.20 \text{ in}^2, \text{ OK}]$$

Use no. 5 at 12 in on center.

The answer is (C).

Why Other Options Are Wrong

(A) This incorrect solution uses the horizontal cross section, instead of the vertical section, in calculating A_g.

(B) This incorrect solution uses the minimum longitudinal reinforcement ratio instead of the transverse ratio.

(D) This incorrect solution uses the maximum spacing limit which does not satisfy the area of reinforcement requirement.

SOLUTION 44

The ratio of vertical shear reinforcement area to gross concrete area of a horizontal section for a shear wall is given in ACI 318 as the larger of 0.0025 and

$$\rho_l = 0.0025 + (0.5)\left(2.5 - \frac{h_w}{l_w}\right)(\rho_t - 0.0025) \quad \text{[ACI 318 Eq. 11.6.2]}$$

Per ACI Sec. 11.6.2, this value need not exceed the required horizontal shear reinforcement, ρ_t, which is calculated per ACI Sec. 11.5.4.8.

$$h_w = (10 \text{ ft})\left(12 \ \frac{\text{in}}{\text{ft}}\right) = 120 \text{ in}$$

$$l_w = (30 \text{ ft})\left(12 \ \frac{\text{in}}{\text{ft}}\right) = 360 \text{ in}$$

$$\rho_l = 0.0025 + (0.5)\left(2.5 - \frac{h_w}{l_w}\right)(\rho_t - 0.0025)$$

$$= 0.0025 + (0.5)\left(2.5 - \frac{120 \text{ in}}{360 \text{ in}}\right)(0.0040 - 0.0025)$$

$$= 0.00413$$

$$\rho_t = 0.0040$$

Since ρ_l need not be greater than ρ_t, ρ_l is 0.0040.

The answer is (C).

Why Other Options Are Wrong

(A) This incorrect solution reverses the height and length of the wall in ACI 318 Eq. 11.6.2 and ignores the minimum reinforcement requirement of ACI 318 Sec. 11.6.2.

(B) This incorrect solution reverses the height and length of the wall in ACI 318 Eq. 11.6.2.

(D) This incorrect solution correctly calculates the vertical shear reinforcement ratio but does not choose the lesser value as given in ACI 318 Sec. 11.6.2.

SOLUTION 45

From a foundation design reference, the minimum depth of penetration, D_{min}, for the sheet piling shown is given by the equation

$$D_{min}^4 - \left(\frac{8H}{\gamma(k_p - k_a)b}\right)D_{min}^2 - \left(\frac{12HL}{\gamma(k_p - k_a)b}\right) \times D_{min} - \left(\frac{2H}{\gamma(k_p - k_a)b}\right)^2 = 0$$

$$k_a = \tan^2\left(45° - \frac{\phi}{2}\right) = \tan^2\left(45° - \frac{30°}{2}\right) = 0.333$$

$$k_p = \tan^2\left(45° + \frac{\phi}{2}\right) = \tan^2\left(45° + \frac{30°}{2}\right) = 3.00$$

$$\gamma = 110 \text{ lbf/ft}^3$$

From the problem statement, the width of the sheet piling, b, is 2 ft; the single concentrated load, H, is 10 kips (10,000 lbf); and the length of the sheet piling above grade, L, is 10 ft.

$$D_{min}^4 - \left(\frac{(8)(10{,}000 \text{ lbf})}{\left(110 \frac{\text{lbf}}{\text{ft}^3}\right)(3.00 - 0.333)(2 \text{ ft})}\right)D_{min}^2 - \left(\frac{(12)(10{,}000 \text{ lbf})(10 \text{ ft})}{\left(110 \frac{\text{lbf}}{\text{ft}^3}\right)(3.00 - 0.333)(2 \text{ ft})}\right)D_{min} - \left(\frac{(2)(10{,}000 \text{ lbf})}{\left(110 \frac{\text{lbf}}{\text{ft}^3}\right)(3.00 - 0.333)(2 \text{ ft})}\right)^2 = 0$$

$$D_{min}^4 - (136.3 \text{ ft}^2)D_{min}^2 - (2045.2 \text{ ft}^3)D_{min} - 1161.9 \text{ ft}^4 = 0$$

This equation can be solved iteratively.

Try $D_{min} = 10$ ft.

$$(10 \text{ ft})^4 - (136.3 \text{ ft}^2)(10 \text{ ft})^2 - (2045.2 \text{ ft}^3)(10 \text{ ft}) - 1161.9 \text{ ft}^4 = 0$$
$$-25{,}244 \text{ ft}^4 \neq 0$$

Try $D_{min} = 15$ ft.

$$(15 \text{ ft})^4 - (136.3 \text{ ft}^2)(15 \text{ ft})^2 - (2045.2 \text{ ft}^3)(15 \text{ ft}) - 1161.9 \text{ ft}^4 = 0$$
$$-11{,}882 \text{ ft}^4 \neq 0$$

Try $D_{min} = 16$ ft.

$$(16 \text{ ft})^4 - (136.3 \text{ ft}^2)(16 \text{ ft})^2 - (2045.2 \text{ ft}^3)(16 \text{ ft}) - 1161.9 \text{ ft}^4 = 0$$
$$-3242 \text{ ft}^4 \neq 0$$

Try $D_{min} = 16.3$ ft.

$$(16.3 \text{ ft})^4 - (136.3 \text{ ft}^2)(16.3 \text{ ft})^2 - (2045.2 \text{ ft}^3)(16.3 \text{ ft}) - 1161.9 \text{ ft}^4 = 0$$
$$-121 \text{ ft}^4 \approx 0$$

This last solution is close to equaling zero and is precise enough for this example. Further iterations would yield an exact solution.

The minimum depth of penetration is

$$D_{min} = 16.3 \text{ ft}$$

Applying the factor of safety of 1.3 yields

$$D = (1.3)(16.3 \text{ ft}) = 21.2 \text{ ft}$$

The total length of the sheet piling is

$$L_{total} = L + D = 10 \text{ ft} + 21.2 \text{ ft} = 31.2 \text{ ft} \quad (31 \text{ ft})$$

The answer is (C).

Why Other Options Are Wrong

(A) This incorrect solution calculates the depth of penetration only, not the total length of the pile.

(B) This incorrect solution does not apply the factor of safety when calculating the required depth.

(D) This incorrect solution does not include the width of the sheet piling, b, in the equation for calculating the minimum depth of penetration, D_{min}.

SOLUTION 46

Chapter 6 of the NDS contains the specifications for round timber piles. Using ASD, the allowable compression design value is

$$F_c' = F_c C_D C_t C_{ct} C_p C_{cs} C_{ls}$$

The reference compression design values, F_c, are based on single piles. According to NDS Sec. 6.3.11, the load sharing factor, C_{ls}, for a group of three piles is 1.09.

From NDS Table 2.3.3, for in-service temperatures less than 100°F,

$$C_t = 1.0$$

From NDS Table 6.3.5, for kiln-dried piles made of red pine,

$$C_{ct} = 0.90$$
$$C_p = 0.62 \quad \text{[given]}$$

Since no information is available regarding the critical section location, use

$$C_{cs} = 1.0$$
$$F_c = 850 \text{ lbf/in}^2 \quad \text{[NDS Table 6A]}$$

The load duration factor depends upon the type of load. The load duration factor for the shortest duration load in a combination of loads shall apply for that load combination [NDS Sec. 2.3.2.2]. There are two load cases to consider: dead load and dead load plus live load.

Dead Load Only:

$$\begin{aligned} C_D &= 0.9 \\ F_c' &= F_c C_D C_t C_{ct} C_p C_{cs} C_{ls} \\ &= \left(850 \ \frac{\text{lbf}}{\text{in}^2}\right)(0.9)(1.0)(0.90)(0.62)(1.0)(1.09) \\ &= 465 \text{ lbf/in}^2 \end{aligned}$$

Dead Load Plus Live Load:

$$\begin{aligned} C_D &= 1.0 \\ F_c' &= F_c C_D C_t C_{ct} C_p C_{cs} C_{ls} \\ &= \left(850 \ \frac{\text{lbf}}{\text{in}^2}\right)(1.0)(1.0)(0.90)(0.62)(1.0)(1.09) \\ &= 517 \text{ lbf/in}^2 \end{aligned}$$

Determine the critical load case.

Dead Load Only:

$$\begin{aligned} f_{c,D} = \frac{P_D}{A} &= \frac{\left(\frac{300 \text{ kips}}{3 \text{ piles}}\right)\left(1000 \ \frac{\text{lbf}}{\text{kip}}\right)}{230 \text{ in}^2} \\ &= 435 \text{ lbf/in}^2 \text{ per pile} \quad [< F_c', \text{OK}] \end{aligned}$$

Dead Load Plus Live Load:

$$\begin{aligned} f_{c,D+L} &= \frac{P_{D+L}}{A} \\ &= \frac{\left(\frac{300 \text{ kips} + 400 \text{ kips}}{3 \text{ piles}}\right)\left(1000 \ \frac{\text{lbf}}{\text{kip}}\right)}{230 \text{ in}^2} \\ &= 1014 \text{ lbf/in}^2 \text{ per pile} \quad [> F_c', \text{NG}] \end{aligned}$$

The critical load case is dead load plus live load. The total adjustment factor for this load case is

$$\begin{aligned} C_D C_t C_{ct} C_p C_{cs} C_{ls} &= (1.0)(1.0)(0.90)(0.62)(1.0)(1.09) \\ &= 0.61 \end{aligned}$$

The answer is (C).

Why Other Options Are Wrong

(A) This incorrect solution does not apply the load sharing factor, C_{ls}.

(B) This incorrect solution uses the *smallest* load duration factor (0.9) for the dead plus live load combination instead of using the load duration factor for the *shortest* load (1.0).

(D) This incorrect solution does not include the condition treatment factor, C_{ct}.

SOLUTION 47

The sliding force is resisted by friction and adhesion between the soil and the base. Passive restraint from the soil is only considered if it will always be there. In most cases, it is neglected. The sliding force is the sum of the pressure due to the earth and the pressure due to the surcharge. From the *Structural Engineering Reference Manual* or another reference book,

$$R_{a,h} = \tfrac{1}{2}k_a\gamma H^2 + k_a qH$$

The lateral earth pressure due to the soil backfill is

$$\begin{aligned} R_{a,h,\text{soil}} = \tfrac{1}{2}k_a\gamma H^2 &= \left(\frac{1}{2}\right)(0.361)\left(0.110 \ \frac{\text{kip}}{\text{ft}^3}\right)(20 \text{ ft})^2 \\ &= 7.94 \text{ kips/ft} \end{aligned}$$

The lateral earth pressure due to the surcharge is

$$\begin{aligned} R_{a,h,\text{surcharge}} &= k_a qH \\ &= (0.361)\left(0.125 \ \frac{\text{kip}}{\text{ft}^2}\right)(20 \text{ ft}) \\ &= 0.90 \text{ kip/ft} \end{aligned}$$

$$R_{a,h} = \tfrac{1}{2}k_a\gamma H^2 + k_a qH = 7.94\ \frac{\text{kips}}{\text{ft}} + 0.90\ \frac{\text{kip}}{\text{ft}}$$
$$= 8.84\ \text{kips/ft}$$

Sliding is resisted by friction from the weight of the soil, retaining wall, and surcharge. From a reference handbook such as the *Structural Engineering Reference Manual*,

$$R_{\text{SL}} = \left(\sum W_i + R_{a,v}\right)\tan\delta + c_A B$$
$$W_i = \gamma_i A_i$$

Since the backfill has no slope, there is no vertical component of the active earth pressure.

$$R_{a,v} = 0$$

According to IBC Table 1806.2, without other information, $\tan\delta$ should be taken as 0.25 for sandy soil without silt.

$$c_A = 0 \quad \text{[for granular soil]}$$
$$R_{\text{SL}} = \left(\sum W_i\right)\tan\delta = \left(\sum W_i\right)(0.25)$$

Determine the weight of the soil, wall, and surcharge.

Soil:

$$A_{\text{soil}} = LH = \left(15\ \text{ft} - \left(\frac{20\ \text{in}}{12\ \frac{\text{in}}{\text{ft}}}\right)\right)(20\ \text{ft} - 2\ \text{ft})$$
$$= 240\ \text{ft}^2$$
$$W_{\text{soil}} = \gamma_{\text{soil}}A_{\text{soil}} = \left(0.110\ \frac{\text{kip}}{\text{ft}^3}\right)(240\ \text{ft}^2)$$
$$= 26.4\ \text{kips/ft}$$

Retaining Wall, Vertical Section:

$$A_{w,v} = LH = \left(\frac{20\ \text{in}}{12\ \frac{\text{in}}{\text{ft}}}\right)(20\ \text{ft}) = 33.3\ \text{ft}^2$$
$$W_{w,v} = \gamma_w A_w = \left(0.150\ \frac{\text{kip}}{\text{ft}^3}\right)(33.3\ \text{ft}^2) = 5.00\ \text{kips/ft}$$

Retaining Wall, Horizontal Section:

$$A_{w,h} = LH = (13.33\ \text{ft})(2\ \text{ft})$$
$$= 26.7\ \text{ft}^2$$
$$W_{w,h} = \gamma_w A_w = \left(0.150\ \frac{\text{kip}}{\text{ft}^3}\right)(26.7\ \text{ft}^2)$$
$$= 4.01\ \text{kips/ft}$$

Surcharge:

$$W_{\text{surcharge}} = \left(0.125\ \frac{\text{kip}}{\text{ft}^2}\right)(8\ \text{ft}) = 1.0\ \text{kip/ft}$$

$$\sum W_i = W_{\text{soil}} + W_{w,v} + W_{w,h} + W_{\text{surcharge}}$$
$$= 26.4\ \frac{\text{kips}}{\text{ft}} + 5.00\ \frac{\text{kips}}{\text{ft}} + 4.01\ \frac{\text{kips}}{\text{ft}} + 1.0\ \frac{\text{kip}}{\text{ft}}$$
$$= 36.4\ \text{kips/ft}$$

The resistance against sliding is

$$R_{\text{SL}} = \left(\sum W_i\right)\tan\delta = \left(36.4\ \frac{\text{kips}}{\text{ft}}\right)(0.25)$$
$$= 9.1\ \text{kips/ft}$$

The factor of safety against sliding is

$$F_{\text{SL}} = \frac{R_{\text{SL}}}{R_{a,h}} = \frac{9.1\ \frac{\text{kips}}{\text{ft}}}{8.84\ \frac{\text{kips}}{\text{ft}}} = 1.03$$

The answer is (A).

Why Other Options Are Wrong

(B) This incorrect solution neglects the effect of the surcharge entirely.

(C) In this incorrect solution, the coefficient of friction is taken as 0.35 instead of 0.25.

(D) This incorrect solution includes the restraint provided by the passive force. Unless explicitly stated that it will always be there, passive restraint should be ignored.

SOLUTION 48

Since the compressive strength of masonry is not specified and material properties are not given, consider empirical design. Appendix A of *Building Code Requirements for Masonry Structures* (TMS 402) contains the empirical requirements for thickness. Two requirements must be checked: one for lateral support and one for minimum thickness. The greater value applies.

TMS 402/602 Table A.5.1 gives wall lateral support requirements. Fully grouted bearing walls have a maximum length-to-thickness ratio, l/t, or height-to-thickness ratio,

h/t, of 20. The limiting length or height is the shortest distance between lateral supports. If the wall spans horizontally, the limiting ratio is the length-to-thickness, l/t. If the wall spans vertically, the limiting ratio is the height-to-thickness, h/t. The wall will span in the shorter direction, which is vertically in this case.

Since the wall spans vertically, $h/t \leq 20$.

Solving for the thickness,

$$t \geq \frac{h}{20} = \frac{(10\ \text{ft})\left(12\ \frac{\text{in}}{\text{ft}}\right)}{20} = 6.0\ \text{in}$$

The minimum thickness requirements of TMS 402 Sec. A.6.2 also apply. This section specifies the minimum wall thickness based on the number of stories. Since the problem states that the wall spans 10 ft from the foundation to the roof, assume the wall is one story. One-story walls must have a minimum thickness of 6 in.

The answer is (A).

Why Other Options Are Wrong

(B) This incorrect solution does not recognize the wall described as a single story and uses the minimum thickness for walls more than one story (8 in).

(C) This incorrect solution uses the wall's length-to-thickness ratio as the limiting ratio. Although lateral support is provided by the intersecting walls, the wall will span the shortest direction, vertically. The span supports at the foundation and roof provide lateral support as well. The limiting ratio should be based on the direction of span, h/t.

(D) This incorrect solution does not recognize the wall described as being an empirically designed wall. Walls designed using allowable stress design (ASD) do not have limits on their thicknesses.

The fact that the compressive strength of masonry, f'_m, is not specified identifies this wall as empirically designed. Walls designed using allowable stress design must specify f'_m.

SOLUTION 49

The first part of the designation, 20F, indicates the tensile capacity in bending, which is 2000 lbf/in^2.

The meaning of the combination symbols for glulam timbers can be found in *Standard Specification for Structural Glued Laminated Timber of Softwood Species* (AITC 117) and in references on wood and timber design.

The answer is (D).

Why Other Options Are Wrong

(A) This answer is incorrect. The "V" in the combination symbol indicates that the glulam beam is visually graded. An "E" signifies mechanically graded beams.

(B) This answer is incorrect. The depth of the beam is as designed and is not indicated in the combination symbol.

(C) This answer incorrectly assumes the shear capacity is indicated by V7. The "V" indicates that the member is visually graded. The "7" is part of the combination symbol and is not an indication of strength.

SOLUTION 50

NDS Sec. 15.3.2 contains the equations for calculating the column stability factor for built-up columns. The column stability factor is based on the slenderness ratios.

First, determine the dimensional properties of the column.

From NDS Supplement Table 1B, the actual dimensions of 2 × 6 sawn lumber are 1.5 in × 5.5 in.

Using NDS Fig. 15B,

$$d_1 = 5.5\ \text{in}$$

$$l_1 = (9\ \text{ft})\left(12\ \frac{\text{in}}{\text{ft}}\right) = 108\ \text{in}$$

The column stability factor, C_p, is based on the effective column length, l_e.

$$l_e = K_e l \quad \text{[NDS Sec. 15.3.2.1]}$$

From Table G1 in App. G of the NDS, the buckling length coefficient, K_e, is 1.0 for a column with both ends free to rotate but not free to translate.

$$l_{e1} = K_e l_1 = (1.0)(108\ \text{in}) = 108\ \text{in}$$

Calculate the slenderness ratio. NDS Sec. 15.3.2.3 specifies that the slenderness ratios shall not exceed 50. The slenderness ratio in direction d_1 is

$$\frac{l_{e1}}{d_1} = \frac{108\ \text{in}}{5.5\ \text{in}} = 19.6 \quad [<50,\ \text{OK}]$$

Using NDS Eq. 15.3-1, calculate the column stability factor in direction d_1.

$$C_p = K_f\left(\frac{1+\frac{F_{c,E}}{F_c^*}}{2c} - \sqrt{\left(\frac{1+\frac{F_{c,E}}{F_c^*}}{2c}\right)^2 - \frac{\frac{F_{c,E}}{F_c^*}}{c}}\right)$$

$$F_c^* = 1750 \text{ lbf/in}^2 \quad \text{[given]}$$

$$F_{c,E} = \frac{0.822E'_{\text{min}}}{\left(\frac{l_e}{d}\right)^2}$$

$$E'_{\text{min}} = 620{,}000 \text{ lbf/in}^2 \quad \text{[given]}$$

$$c = 0.8 \quad \text{[for sawn lumber]}$$

$$\frac{l_{e1}}{d_1} = 19.6$$

$$K_f = 1.0 \quad \left[\begin{array}{c}\text{for a built-up column where } l_{e1}/d_1 \\ \text{is used to calculate } F_{c,E}\end{array}\right]$$

$$F_{c,E_1} = \frac{0.822E'_{\text{min}}}{\left(\frac{l_{e1}}{d_1}\right)^2} = \frac{(0.822)\left(620{,}000 \ \frac{\text{lbf}}{\text{in}^2}\right)}{(19.6)^2}$$
$$= 1327 \text{ lbf/in}^2$$

$$\frac{F_{c,E_1}}{F_c^*} = \frac{1327 \ \frac{\text{lbf}}{\text{in}^2}}{1750 \ \frac{\text{lbf}}{\text{in}^2}} = 0.758$$

$$C_{p1} = K_f\left(\frac{1+\frac{F_{c,E}}{F_c^*}}{2c} - \sqrt{\left(\frac{1+\frac{F_{c,E}}{F_c^*}}{2c}\right)^2 - \frac{\frac{F_{c,E}}{F_c^*}}{c}}\right)$$
$$= (1.0)\left(\frac{1+0.758}{(2)(0.8)} - \sqrt{\left(\frac{1+0.758}{(2)(0.8)}\right)^2 - \frac{0.758}{0.8}}\right)$$
$$= 0.59$$

The answer is (C).

Why Other Options Are Wrong

(A) This incorrect solution uses a buckling length coefficient, K_e, of 0.65 from NDS Table G1 in App. G for a column with rotation and translation fixed at both ends.

(B) This incorrect solution finds the column stability factor in direction 2 instead of direction 1.

(D) This incorrect solution calculates the slenderness ratio based on the nominal dimensions of the wood (2 in × 6 in), rather than using the actual dimensions.

SOLUTION 51

ASTM specification C652 is the *Standard Specification for Hollow Brick (Hollow Masonry Units Made from Clay or Shale)*. Therefore, the equation for the modulus of elasticity for clay masonry must be used.

The modulus of elasticity of clay masonry is given in *Building Code Requirements for Masonry Structures* (TMS 402) Sec. 4.2.2.2.1 as

$$E_m = 700f'_m = (700)\left(4950 \ \frac{\text{lbf}}{\text{in}^2}\right)$$
$$= 3{,}465{,}000 \text{ lbf/in}^2 \quad (3.5 \times 10^6 \text{ lbf/in}^2)$$

The answer is (C).

Why Other Options Are Wrong

(A) This incorrect solution misreads the specified compressive strength of masonry given in the problem statement as the compressive strength of the masonry unit. It uses Part 1, Table 1 of the *Specification for Masonry Structures* (TMS 602) to determine the compressive strength of 4950 lbf/in^2 clay masonry laid in Type S mortar to be 2000 lbf/in^2 and uses this value in calculating the modulus of elasticity.

(B) This incorrect solution mistakenly identifies that the unit is made from concrete and misreads the specified compressive strength of masonry given in the problem statement as the compressive strength of the masonry unit. It uses TMS 602 Part 1, Table 2 and the equation for modulus of elasticity, E_m, for concrete masonry units.

(D) This incorrect solution uses the equation for the modulus of elasticity of concrete masonry instead of the equation for clay masonry.

SOLUTION 52

Statement I is true. Mat foundations are useful in areas where the basement is below the ground water table (GWT) because the mat foundation provides a water barrier, since it is a continuous concrete slab.

Statement II is true. Mat foundations require both top and bottom reinforcement because both positive and negative moments are developed in the foundation.

The answer is (B).

Why Other Options Are Wrong

(A) This incorrect solution only identifies statement I as true and misses the fact that statement II is also true.

(C) This incorrect solution mistakenly identifies the two solutions that are false rather than those that are true. Statements III and IV are false. Mat foundations are suitable where settlements may be a problem or where settlements may be large due, in part, to their rigidity. The rigidity of the mat tends to bridge over areas of erratic soil types or voids.

(D) This incorrect solution wrongly identifies statements III and IV as true. Mat foundations are suitable where settlements may be a problem or where settlements may be large due, in part, to their rigidity. The rigidity of the mat tends to bridge over areas of variable soil types or voids. Statements III and IV are false.

SOLUTION 53

Determine the allowable loads using *Building Code Requirements for Masonry Structures* (TMS 402). The allowable load in shear, B_v, of headed anchor bolts is the smallest of the allowable shear loads governed by masonry breakout (TMS 402 Eq. 8-6), by masonry crushing (TMS 402 Eq. 8-7), by anchor bolt pryout (TMS 402 Eq. 8-8), and by steel yielding (TMS 402 Eq. 8-9). $A_{p,v}$ is the projected area for shear and is calculated from the anchor bolt edge distance in the direction of the load, l_{be}, and TMS 402 Eq. 6-2. Since edge distance is not a concern, $A_{p,v}$ will not control. Therefore, $B_{v,b}$ can be disregarded.

First, calculate the projected area for axial tension, $A_{p,t}$. The effective embedment length, l_b, is 6 in.

$$\begin{aligned} A_{p,t} &= \pi l_b^2 = \pi (6 \text{ in})^2 \\ &= 113 \text{ in}^2 \quad \text{[TMS 402 Eq. 6-1]} \end{aligned}$$

Calculate the allowable shear loads. A ¾ in diameter bolt has an area, A_b, of 0.44 in^2.

$$\begin{aligned} B_{v,c} &= 350\sqrt[4]{f'_m A_b} \quad \text{[TMS 402 Eq. 8-7]} \\ &= 350\sqrt[4]{\left(2000 \frac{\text{lbf}}{\text{in}^2}\right)(0.44 \text{ in}^2)} \\ &= 1906 \text{ lbf} \quad (1900 \text{ lbf}) \quad \text{[controls]} \end{aligned}$$

$$\begin{aligned} B_{v,\text{pry}} &= 2.5A_{p,t}\sqrt{f'_m} \quad \text{[TMS 402 Eq. 8-8]} \\ &= (2.5)(113 \text{ in}^2)\sqrt{2000 \frac{\text{lbf}}{\text{in}^2}} \\ &= 12{,}634 \text{ lbf} \quad (12{,}600 \text{ lbf}) \end{aligned}$$

$$\begin{aligned} B_{v,s} &= 0.36A_b f_y \quad \text{[TMS 402 Eq. 8-9]} \\ &= (0.36)(0.44 \text{ in}^2)\left(36{,}000 \frac{\text{lbf}}{\text{in}^2}\right) \\ &= 5702 \text{ lbf} \quad (5700 \text{ lbf}) \end{aligned}$$

The allowable load in shear is the smallest allowable load. Therefore, masonry crushing controls.

The answer is (B).

Why Other Options Are Wrong

(A) This incorrect solution uses a specified compressive strength of masonry of 1500 lbf/in^2 instead of 2000 lbf/in^2 as given in the problem statement.

(C) This incorrect solution finds the square root instead of the fourth root when calculating the allowable shear load by masonry crushing, making the steel yielding the controlling value.

(D) This incorrect solution uses the maximum value instead of the minimum value as the correct answer.

SOLUTION 54

ACI 318 Sec. 19.3.3.1 specifies that concrete exposed to freezing and thawing conditions (exposure classes F1, F2 or F3) must be air-entrained for durability. The total air content for frost-resistant concrete is given in ACI 318 Table 19.3.3.1 as a function of exposure class and aggregate size. ACI 318 Table 19.3.1.1 and Table R19.3.1 assign exposure class F3 to horizontal members in parking structures. For exposure class F3 and an aggregate size of 1 in, the total air content of the concrete mix should be 6.0%.

The answer is (D).

Why Other Options Are Wrong

(A) In this incorrect solution, the units on the table have been misread, and the value is "converted" to a percentage by dividing by 100.

(B) This incorrect solution finds the correct exposure class, F3, but uses the maximum water-cementitious materials ratio from ACI 318 Table 19.3.2.1 (Requirements for Concrete by Exposure Class) instead of the air content from ACI 318 Table 19.3.3.1.

(C) This incorrect solution uses the value from ACI 318 Table 19.3.3.1 for exposure class F1, rather than for exposure class F3.

SOLUTION 55

The temperature factor, C_t, is discussed in NDS Sec. 2.3.3. The temperature factor applies to members subjected to sustained exposure to elevated (over 100°F) temperatures. Because the statement applies to members subjected to extreme cold, not heat, statement I is false.

The volume factor, C_V, is discussed in NDS Sec. 5.3 on structural glued laminated timber and in Sec. 8.3 on structural composite lumber. Statement II is true.

The repetitive member factor, C_r, discussed in Sec. 4.3.9, applies only to dimension lumber bending members 2 in to 4 in thick. Because the joists in this case are 1 in thick, statement III is false.

The load duration factor, C_D, is discussed in Sec. 2.3.2 and NDS Table 2.3.2. Both NDS Sec. 2.3.2 and Footnote 1 to NDS Table 2.3.2 state that "Load duration factors shall not apply to reference modulus of elasticity, *E*..." Statement IV is true.

The answer is (C).

Why Other Options Are Wrong

(A) Although statement II is true, statement I is false.

(B) Although statement II is true, statement III is false.

(D) Although statement IV is true, statement III is false.

SOLUTION 56

Find the section properties of an ASTM A992 W18 × 40 beam in Part 1 of the AISC *Steel Construction Manual.*

$$b_f = 6.02 \text{ in}$$
$$A = 11.8 \text{ in}^2$$
$$d = 17.9 \text{ in}$$

To find the centroid of the transformed area, determine the equivalent width of the transformed concrete.

$$b_{\text{transformed}} = \frac{1}{n}b_e = \left(\frac{1}{8}\right)(72 \text{ in}) = 9.0 \text{ in}$$

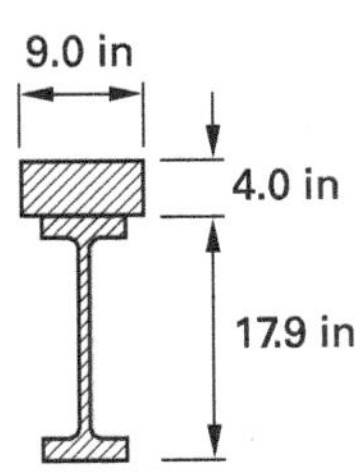

The distance to the centroid of the transformed section measured from the bottom of the beam is

$$\bar{y} = \frac{\sum A_i \bar{y}}{\sum A_i} = \frac{(9.0 \text{ in})(4.0 \text{ in})\left(17.9 \text{ in} + \frac{4.0 \text{ in}}{2}\right) + (11.8 \text{ in}^2)\left(\frac{17.9 \text{ in}}{2}\right)}{(9.0 \text{ in})(4.0 \text{ in}) + 11.8 \text{ in}^2} = 17.2 \text{ in} \quad (17 \text{ in})$$

The answer is (C).

Why Other Options Are Wrong

(A) This incorrect solution makes a math error in calculating the transformed area, subtracting half the thickness of the concrete instead of adding it to find the distance of the transformed concrete.

The distance to the centroid of the transformed section measured from the bottom of the beam is

$$\bar{y} = \frac{\sum A_i \bar{y}}{\sum A_i} = \frac{(9.0 \text{ in})(4.0 \text{ in})\left(17.9 \text{ in} - \frac{4.0 \text{ in}}{2}\right) + (11.8 \text{ in}^2)\left(\frac{17.9 \text{ in}}{2}\right)}{(9.0 \text{ in})(4.0 \text{ in}) + 11.8 \text{ in}^2} = 14.2 \text{ in} \quad (14 \text{ in})$$

(B) This incorrect solution uses the section properties of a W16 × 40 beam instead of a W18 × 40 beam.

(D) This incorrect solution does not transform the concrete when calculating the centroid.

Design & Details of Structures

SOLUTION 57

Since the support condition of the beam is not known, from ACI 318 Sec. 24.5.3.2, the extreme fiber stress in tension in the concrete is limited to

$$\sigma = 3\sqrt{f'_{ci}} = 3\sqrt{4000\ \frac{\text{lbf}}{\text{in}^2}}$$
$$= 189.7\ \text{lbf/in}^2 \quad (190\ \text{lbf/in}^2)$$

The answer is (B).

Why Other Options Are Wrong

(A) This incorrect solution identifies the initial prestress force in the tendon, P_i, as the tensile stress. Note that the units for initial prestress force are actually kips, not lbf/in^2.

(C) This incorrect solution uses specified strength of concrete rather than the concrete strength at the time of initial prestress when calculating the allowable stress.

(D) This incorrect solution uses the stress limit for simply supported members found in ACI 318 Table 24.5.3.2.